LONDON MATHEMATICAL SOCIETY LECTURE NOTE SERIES

Managing Editor: Professor I.M.James,
Mathematical Institute, 24-29 St Giles, Oxford

1. General cohomology theory and K-theory, P.HILTON
4. Algebraic topology: a student's guide, J.F.ADAMS
5. Commutative algebra, J.T.KNIGHT
8. Integration and harmonic analysis on compact groups, R.E.EDWARDS
9. Elliptic functions and elliptic curves, P.DU VAL
10. Numerical ranges II, F.F.BONSALL & J.DUNCAN
11. New developments in topology, G.SEGAL (ed.)
12. Symposium on complex analysis, Canterbury, 1973, J.CLUNIE & W.K.HAYMAN (eds.)
13. Combinatorics: Proceedings of the British combinatorial conference 1973, T.P.McDONOUGH & V.C.MAVRON (eds.)
14. Analytic theory of abelian varieties, H.P.F.SWINNERTON-DYER
15. An introduction to topological groups, P.J.HIGGINS
16. Topics in finite groups, T.M.GAGEN
17. Differentiable germs and catastrophes, Th.BROCKER & L.LANDER
18. A geometric approach to homology theory, S.BUONCRISTIANO, C.P.ROURKE & B.J.SANDERSON
20. Sheaf theory, B.R.TENNISON
21. Automatic continuity of linear operators, A.M.SINCLAIR
23. Parallelisms of complete designs, P.J.CAMERON
24. The topology of Stiefel manifolds, I.M.JAMES
25. Lie groups and compact groups, J.F.PRICE
26. Transformation groups: Proceedings of the conference in the University of Newcastle upon Tyne, August 1976, C.KOSNIOWSKI
27. Skew field constructions, P.M.COHN
28. Brownian motion, Hardy spaces and bounded mean oscillation, K.E.PETERSEN
29. Pontryagin duality and the structure of locally compact abelian groups, S.A.MORRIS
30. Interaction models, N.L.BIGGS
31. Continuous crossed products and type III von Neumann algebras, A.VAN DAELE
32. Uniform algebras and Jensen measures, T.W.GAMELIN
33. Permutation groups and combinatorial structures, N.L.BIGGS & A.T.WHITE
34. Representation theory of Lie groups, M.F.ATIYAH et al.
35. Trace ideals and their applications, B.SIMON
36. Homological group theory, C.T.C.WALL (ed.)
37. Partially ordered rings and semi-algebraic geometry, G.W.BRUMFIEL
38. Surveys in combinatorics, B.BOLLOBAS (ed.)
39. Affine sets and affine groups, D.G.NORTHCOTT
40. Introduction to H_p spaces, P.J.KOOSIS
41. Theory and applications of Hopf bifurcation, B.D.HASSARD, N.D.KAZARINOFF & Y-H.WAN
42. Topics in the theory of group presentations, D.L.JOHNSON
43. Graphs, codes and designs, P.J.CAMERON & J.H.VAN LINT
44. Z/2-homotopy theory, M.C.CRABB
45. Recursion theory: its generalisations and applications, F.R.DRAKE & S.S.WAINER (eds.)
46. p-adic analysis: a short course on recent work, N.KOBLITZ
47. Coding the Universe, A. BELLER, R. JENSEN & P. WELCH
48. Low-dimensional topology, R. BROWN & T.L. THICKSTUN (eds.)
49. Finite geometries and designs, P. CAMERON, J.W.P. HIRSCHFELD & D.R. HUGHES (eds.)
50. Commutator Calculus and groups of homotopy classes, H.J. BAUES
51. Synthetic differential geometry, A. KOCK
52. Combinatorics, H.N.V. TEMPERLEY (ed.)
53. Singularity theory, V.I. ARNOLD
54. Markov processes and related problems of analysis, E.B. DYNKIN
55. Ordered permutation groups, A.M.W. GLASS
56. Journees arithmetiques 1980, J.V. ARMITAGE (ed.)
57. Techniques of geometric topology, R.A. FENN
58. Singularities of differentiable functions, J. MARTINET
59. Applicable differential geometry, F.A.E. PIRANI and M. CRAMPIN
60. Integrable systems, S.P. NOVIKOV et al.

61. The core model, A. DODD
62. Economics for mathematicians, J.W.S. CASSELS
63. Continuous semigroups in Banach algebras, A.M. SINCLAIR
64. Basic concepts of enriched category theory, G.M. KELLY

London Mathematical Society Lecture Note Series. 62

Economics for Mathematicians

J.W.S. CASSELS

CAMBRIDGE UNIVERSITY PRESS

CAMBRIDGE

LONDON NEW YORK NEW ROCHELLE

MELBOURNE SYDNEY

Published by the Press Syndicate of the University of Cambridge
The Pitt Building, Trumpington Street, Cambridge CB2 1RP
32 East 57th Street, New York, NY 10022, USA
296 Beaconsfield Parade, Middle Park, Melbourne 3206, Australia

First published in 1981

Printed in Great Britain at the University Press, Cambridge

British Library cataloguing in publication data

Cassels, J.W.S.
Economics for Mathematicians.-(London Mathematical Society
Lecture Note Series 62 ISSN 0076-0552.)
1.Economics, Mathematical
I Title.   II Series
330'.02451   HB171.5

ISBN 0 521 28614 X

The construction of economic "models", and indeed the whole of economic theory, can be regarded as a (rather pedestrian) subsection of mathematics, gaining interest only when tested as an explanation of the real world.

Sir Charles Carter. Higher education for the future (Blackwell, Oxford 1980), p.94

But I know I had a growing feeling in the later years of my work on the subject that a good mathematical theorem dealing with economic hypotheses was very unlikely to be good economics: and I went more and more on the rules - (1) Use mathematics as a shorthand language, rather than as an engine of enquiry. (2) Keep to them till you are done. (3) Translate into English. (4) Then illustrate by examples that are important in real life. (5) Burn the mathematics. (6) If you can't succeed in 4, burn 3. This last I did often.

Letter from Marshall to Bowley. Memorials of Alfred Marshall (Ed. A.C. Pigou) (Macmillan, London, 1925), p.427.

CONTENTS

PREFACE

These are the expanded notes of a course intended to introduce mathematicians to some of the central ideas of traditional economics. They are just notes; they lack the "corroborative detail, intended to give artistic verisimilitude to an otherwise bald and unconvincing narrative".

There appears to be no book doing what the course attempts to do. Perhaps the nearest approach is E. Malinvaud, Lectures in microeconomic theory (North Holland and American Elsevier, 1972), which is particularly relevant to the first four chapters. There is also a useful account of the subject matter of Chapters 1-4 in D. Dewey's Microeconomics (O.U.P. paperback, 1975). The topics of Chapter 5 are those of the last chapter of David Gale's admirable Theory of linear economic models (McGraw Hill, 1960), but we take them somewhat further. Finally, the simple models of Chapter 6 are discussed at exhaustive length in P.A. Samuelson's Economics (McGraw Hill, Kogakusha, 10th ed., 1976) (especially Chapters 12,13,18). This is a massive text intended for the mathematically underdeveloped, but it can be read without a shudder. Indeed it seems to be the best general introduction to the background of the whole course and explains the buzzwords, without which no discussion in economics is complete. Adam Smith's Wealth of Nations still makes interesting reading.

There is a fair number of books on the market with the title "Mathematical economics" or something similar. Those I have sampled have been disappointing. They devote considerable space to expounding standard mathematics. When they get down to use it, they tend to be clumsy, and some are not above a blatant fudge.

I am grateful to colleagues and friends, in particular to Dr. H.M. Pesaran, for helpful comments and suggestions. Needless to say, any remaining mistakes and misconceptions are mine alone.

Further reading (in addition to books named above):

R.G.D. Allen. Macroeconomic theory (Macmillan, 1967). Gives a description of a wide zoo of macroeconomic models that have been proposed in the past (with their sometimes beguiling terminology): they mainly involve differential and difference equations.

K. Arrow and F.H. Hahn. Competitive analysis (Holden-Day; Oliver and Boyd, 1971).

R.L. Crouch. Macroeconomics (Harcourt, Brace, Jovanovich, 1972).

S.J. Turnovsky. Macroeconomic theory and stabilization policy (Cambridge University Press, 1977).

NOTATION

Real vectors are denoted by underlined lower case letters (Latin and, occasionally, Greek) with the convention exemplified by $\underline{x} = (x_1,\ldots,x_n)$. The inner product is written e.g.

$$\underline{px} = p_1x_1 +\ldots+ p_nx_n \ .$$

We do not in the notation distinguish between row and column vectors, since which is which should be clear from the context. In general, vectors of goods ("bundles of commodities") are column vectors and prices are row vectors.

There is the following notation for inequalities between vectors of the same dimension:

$\underline{x} \geq \underline{y}$ means $x_j \geq y_j$ (all j) .

$\underline{x} > \underline{y}$ means $\underline{x} \geq \underline{y}$ but $\underline{x} \neq \underline{y}$.

$\underline{x} >> \underline{y}$ means $x_j > y_j$ (all j) .

Sometimes vectors are labelled with suffixes or superfixes e.g. $\underline{x}_i$ or $\underline{y}^i$. In this case the j-th coordinate is x_{ij} or y^i_j respectively.

Partial derivatives may be denoted by suffixes. If $f = f(x_1,\ldots,x_n)$, then $f_j = \partial f/\partial x_j$. Sometimes, however, suffixes are used merely to label functions. Here, again, further suffixes may denote partial derivatives. Thus if f_i $(1 \leq i \leq m)$ is a set of functions of $x_1,\ldots,x_n$, we may write $f_{ij} = \partial f_i/\partial x_j$. The context will make everything crystal clear.

We denote the unit matrix (with 1 on the diagonal and 0 elsewhere) by I and the zero matrix (all of whose elements are 0) by O .

CHAPTER 1. UTILITY, INDIFFERENCE HYPERSURFACES

1. Preliminaries

We shall be modelling the preferences of an individual between "bundles of commodities". We suppose that there are n commodities (or goods) labelled $1,\dots,n$. A bundle of commodities is a real vector $\underline{x} = (x_1,\dots,x_n) \geq \underline{0}$, where x_j is the quantity of commodity j (in some given units). The commodities are supposed to be infinitely divisible, so that the x_j can take all non-negative values.

A given individual is supposed to have an order of preference between any two bundles $\underline{x},\underline{y}$. Either he prefers $\underline{x}$ to $\underline{y}$ (written $\underline{x} \succ \underline{y}$) or he prefers $\underline{y}$ to $\underline{x}$ ($\underline{y} \succ \underline{x}$) or he is indifferent between them (written $\underline{x} \asymp \underline{y}$). If either $\underline{x} \succ \underline{y}$ or $\underline{x} \asymp \underline{y}$ we write $\underline{x} \succsim \underline{y}$. We suppose that the preferences are consistent, in the sense that $\succ$ satisfies the usual postulates for a (pre-) order:

$$\underline{x} \succsim \underline{y} \text{ and } \underline{y} \succsim \underline{x} \Rightarrow \underline{x} \asymp \underline{y} \tag{1.1}$$

$$\underline{x} \succsim \underline{y} \text{ and } \underline{y} \succ \underline{z} \Rightarrow \underline{x} \succ \underline{z} \tag{1.2}$$

$$\underline{x} \succ \underline{y} \text{ and } \underline{y} \succsim \underline{z} \Rightarrow \underline{x} \succ \underline{z} \quad . \tag{1.3}$$

We shall suppose that our individual prefers to have more of each good rather than less (the goods actually are "goods" and not "bads"):

$$\underline{x} \geq \underline{y} \Rightarrow \underline{x} \succsim \underline{y} \quad . \tag{1.4}$$

We shall usually make the stronger supposition that

$$\underline{x} > \underline{y} \Rightarrow \underline{x} \succ \underline{y} \, , \tag{1.5}$$

at least when

$$\underline{y} >> \underline{0} \; . \tag{1.6}$$

The bundles of goods $\underline{y}$ in which a component is 0 may behave anomalously in the theory and we may not always discuss in detail the modifications required in the theory to deal with them.

It is a standard assumption in economics that for any $\underline{x}_o$ the set

$$V(\underline{x}_o) = \{\underline{x} : \underline{x} \succsim \underline{x}_o\} \tag{1.7}$$

is convex. This may be regarded as a consequence of the "law of diminishing returns". The assumption is crucial in most of what follows. There will, however, be one or two places where we shall drop the condition that $V(\underline{x}_o)$ is convex, but then we shall do so explicitly.

We shall be concerned only with preferences that can be given by a <u>utility function</u>. This is a continuous real-valued function $u(\underline{x})$ defined on the bundles of commodities $\underline{x} \geq \underline{0}$. The corresponding preference $\succ$ is given by

$$\underline{x} \succ \underline{y} \iff u(\underline{x}) > u(\underline{y}) .$$

We shall suppose that the $u(\underline{x})$ are continuously differentiable to the extent that the argument requires.

The condition (1.5) implies that the set

$$u(\underline{x}) = \text{constant} \tag{1.8}$$

is a hypersurface. It is called an <u>indifference hypersurface</u>. If $\underline{x}, \underline{y}$ are on the same indifference hypersurface, then $\underline{x} \sim \underline{y}$: and conversely. The convexity assumption on $V(\underline{x}_o)$ implies that indifference hypersurfaces are convex. We shall normally suppose that they are <u>strictly</u> convex.

Commonly used examples of utility functions are

$$\Pi x_j \qquad (1 \leq j \leq n) , \tag{1.9}$$

or, more generally,

$$\Pi x_j^{\alpha(j)} \qquad (1 \leq j \leq n) , \tag{1.10}$$

where the $\alpha(j) > 0$ are constants. These satisfy all the assumptions made so far, except that the indifference hypersurfaces $u(\underline{x}) = 0$ are not strictly convex.

Two utility functions $u(\underline{x})$, $v(\underline{x})$ give the same preference relation if and only if there is a continuous strictly increasing function ϕ such that

$$v(\underline{x}) = \phi(u(\underline{x})) \quad . \tag{1.11}$$

The theory will not distinguish between u and v. That is to say, we shall attach significance to inequalities between $u(\underline{x})$ for different bundles $\underline{x}$ but we shall not attach any meaning to the value $u(\underline{x})$ itself. [Our utilities are ordinal, not cardinal. There are theories with a cardinal utility. Edgeworth constructed the first in his Mathematical Psychics on the basis of the "hedonistic calculus", and, more recently, von Neumann and Morgenstern used the theory of games: but we shall never be concerned with their concepts.]

Warning. Although we suppose that the indifference hypersurfaces are convex, we do NOT suppose that $u(\underline{x})$ is a convex function of $\underline{x}$.

2. Budget constraints

A price vector is a vector $\underline{p} > \underline{0}$ of the same dimensionality n as the bundle of commodities. The j-th component p_j is the price of (a standard unit of) the j-th commodity. If $p_j = 0$, the good j is free. To avoid special cases we shall sometimes assume that no good is free. The price vectors lie in the dual vector space to that of the commodity bundles $\underline{x}$. In particular, the scalar product $\underline{px}$ is defined. It is the cost of $\underline{x}$ at prices $\underline{p}$. [The term "value" is sometimes used, but is usually reserved for the concept introduced in Chapter 5, §3.]

A budget constraint is an inequality of the type

$$\underline{px} \le R \quad , \tag{2.1}$$

where the prices $\underline{p}$ and the budget $R > 0$ are given. An individual maximizes his utility subject to the constraint. Under the conditions of the previous section

Figure 1.

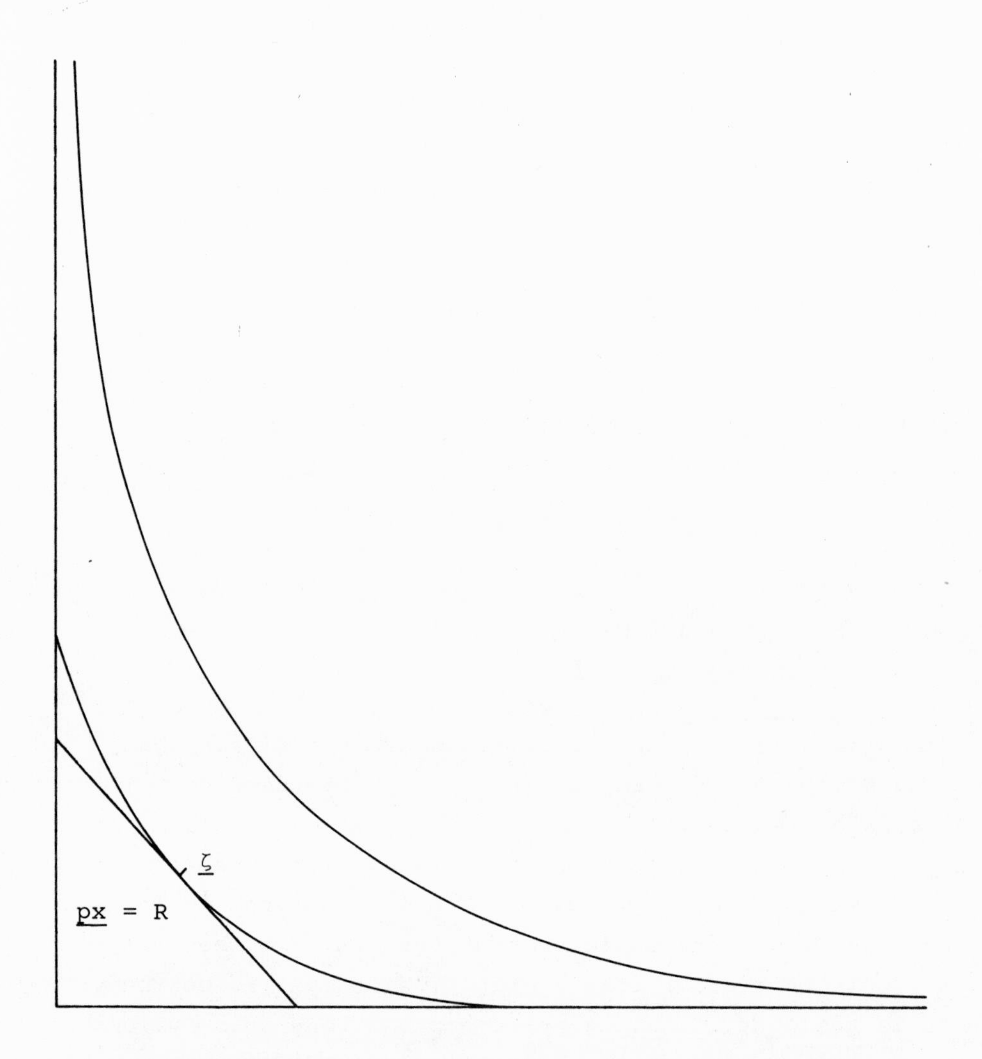

This diagram illustrates (for $n = 2$) choice with a utility function subject to a budget constraint. The point ζ maximizes utility subject to $\underline{p}\underline{x} \leq R$. The line $\underline{p}\underline{x} = R$ is tangent to the indifference curve through ζ. Another indifference curve is also shown.

there is a unique

$$\underline{\zeta} = \underline{\zeta}(\underline{p},R) \qquad (2.2)$$

which does this, and it lies on the hyperplane

$$\underline{px} = R \; . \qquad (2.3)$$

The hyperplane (2.3) is clearly a tac-hyperplane to

$$V(\underline{\zeta}) = \{\underline{x}:u(\underline{x}) \geq u(\underline{\zeta})\} \; . \qquad (2.4)$$

If $\underline{\zeta} >> \underline{0}$, the only tac-hyperplane at $\underline{\zeta}$ is the tangent hyperplane, and so

$$p_j = \lambda u_j(\underline{\zeta}) \qquad (2.5)$$

for some $\lambda > 0$, where by definition,

$$u_j(\underline{x}) = \partial u(\underline{x})/\partial x_j \; . \qquad (2.6)$$

If, however, $\zeta_j = 0$ for some j , then (2.5) need not hold.

Conversely, if $\underline{\zeta}$ is any bundle of commodities with $\underline{\zeta} >> \underline{0}$ and $\underline{p}$ is given by (2.5) for some λ , then $\underline{\zeta} = \underline{\zeta}(\underline{p},R)$ for some budget R .

We shall be examining the behaviour of $\underline{\zeta}(\underline{p},R)$ as $\underline{p}$ and R vary, but first study what turns out to be an easier problem.

3. Indifference hypersurfaces

In this section we consider a single indifference hypersurface U . For every price vector $\underline{p} > \underline{0}$ there is a $\underline{z}(\underline{p}) \in U$ which minimizes the cost $\underline{px}$ for $\underline{x} \in U$. We shall study $\underline{z}(\underline{p})$ as a function of $\underline{p}$. By definition

$$\underline{px} \geq \underline{pz}(\underline{p}) \qquad (\text{all } \underline{x} \in U) \; . \qquad (3.1)$$

On the assumption of strict convexity, there is inequality in (3.1) if $\underline{x} \neq \underline{z}(\underline{p})$.

Now let $\underline{p}^* > \underline{0}$ and $\underline{p}^o > \underline{0}$ be two price vectors and write

$$\underline{z}^* = \underline{z}(\underline{p}^*), \; \underline{z}^o = \underline{z}(\underline{p}^o) \; . \qquad (3.2)$$

By (3.1) we have

$$\underline{p}^o\underline{z}^* \geq \underline{p}^o\underline{z}^o \; ; \; \underline{p}^*\underline{z}^o \geq \underline{p}^*\underline{z}^* \; ; \qquad (3.3)$$

and so

$$(\underline{p}^*-\underline{p}^o)(\underline{z}^*-\underline{z}^o) \leq 0 \; . \qquad (3.4)$$

This is known as <u>the Substitution Theorem</u>.

In particular, if $\underline{p}^*$, $\underline{p}^o$ differ in only one coordinate, say

$$p_1^* > p_1^o \ , \ p_j^* = p_j^o \qquad (j > 1) \ , \tag{3.5}$$

then (3.4) implies that $z_1^* \leq z_1^o$. Hence

$$\partial z_j(\underline{p})/\partial p_j \leq 0 \qquad (1 \leq j \leq n) \ , \tag{3.6}$$

provided that the partial derivative exists.

We can generalize (3.6). Let $\underline{\ell}$ be any vector and put $\underline{p}^o = \underline{p}^* + \delta\underline{\ell}$, where $\delta \to 0$. On substituting in (3.4) we readily obtain

$$\Sigma \ \ell_i \ell_j \partial z_i(\underline{p})/\partial p_j \leq 0 \qquad (\text{all } \underline{\ell}) \ . \tag{3.7}$$

We shall comment on the significance of this below.

Now make an infinitesimal change in price from $\underline{p}$ to $\underline{p} + d\underline{p}$ and let the corresponding change in $\underline{z} = \underline{z}(\underline{p})$ be from $\underline{z}$ to $\underline{z} + d\underline{z}$. Then $\underline{z} + d\underline{z}$ is in the tangent plane to U at $\underline{z}$, and so

$$\underline{p}d\underline{z} = 0 \ . \tag{3.8}$$

We define $r(\underline{p})$ to be the least cost of a bundle on U at prices $\underline{p}$, that is

$$r(\underline{p}) = \underline{p}\underline{z}(\underline{p}) . \tag{3.9}$$

By (3.8) we have

$$\begin{aligned} dr &= \underline{p}d\underline{z} + \underline{z}d\underline{p} \\ &= \underline{z}d\underline{p} \ . \end{aligned} \tag{3.10}$$

Since (3.10) is a perfect differential, we have

$$\partial z_j/\partial p_k = \partial z_k/\partial p_j \ . \tag{3.11}$$

This is the <u>reciprocity theorem</u>.

If the common value of (3.11) is positive, we say that the goods j and k are <u>substitutes</u> (if the price of tea goes up, we drink more coffee, and vice versa). If (3.11) is negative, then j and k are <u>complements</u> (if the price of tea goes up, we use less sugar, and vice versa). By (3.6) and (3.8) we have

$$\begin{aligned} \sum_{j \neq k} p_j(\partial z_j/\partial p_k) &= -p_k(\partial z_k/\partial p_k) \\ &\geq 0 \ , \end{aligned} \tag{3.12}$$

and so in some sense substitutes are more frequent than complements.

Note 1. By definition $\underline{z}(\underline{p})$ is homogeneous of degree 0 in $\underline{p}$:

$$\underline{z}(\lambda\underline{p}) = \underline{z}(\underline{p}) \qquad (\lambda > 0) . \tag{3.13}$$

By Euler's Theorem it follows that

$$\sum_k p_k(\partial z_j/\partial p_k) = 0 \qquad (1 \le j \le n) . \tag{3.14}$$

But this gives no further information, since it follows from (3.8) and (3.11).

Note 2. We do not appear to have used the hypothesis of the convexity of (1.7) except to show that $\underline{z}(\underline{p})$ is uniquely determined by $\underline{p}$, and indeed much of the argument (e.g. the proof of (3.4)) does not even require that. Let us suppose that the indifference hypersurface U is the frontier of (1.7) . If (1.7) is not supposed to be convex, the $\underline{z}(\underline{p})$ will nevertheless all lie on the frontier U^* of the convex cover of (1.7). In other words, if (1.7) is not convex, we should never be able to find this out!

Note 3. The condition (3.7) is essentially equivalent to the definiteness or semi-definiteness of the matrix of second derivative of a convex function (cf Appendix A). For an approach along these lines, see K. Lancaster's Mathematical Economics (not recommended). One defines $\underline{p}$ in terms of $\underline{z}$ instead of vice versa and then uses that $\underline{z}$ runs over a convex hypersurface.

4. Utility functions

We now revert to the situation discussed in section 2. There is a utility function $u(\underline{x})$ and for given price vector $\underline{p} > \underline{0}$ and budget R the vector $\underline{\zeta} = \underline{\zeta}(\underline{p},R)$ maximizes $u(\underline{x})$ subject to $\underline{px} \le R$. We introduce the notation

$$\underline{v} = \partial\underline{\zeta}/\partial R . \tag{4.1}$$

In section 3 we considered an indifference

hypersurface $U : u(\underline{x}) = \text{constant}$. The function $\underline{z}(\underline{p})$ considered there is given by

$$\underline{z}(\underline{p}) = \underline{\zeta}(\underline{p}, r(\underline{p})) , \qquad (4.2)$$

where $r(\underline{p})$ is given by (3.9) . Hence

$$\partial\underline{z}/\partial p_j = \partial\underline{\zeta}/\partial p_j + (\partial\underline{\zeta}/\partial R)(\partial r/\partial p_j)$$

$$= \partial\underline{\zeta}/\partial p_j + \zeta_j\underline{v} \qquad (4.3)$$

by (3.10). It follows that

$$d\underline{\zeta} = Vd\underline{p} + \underline{v}(dR - \underline{\zeta}d\underline{p}) , \qquad (4.4)$$

where V is the matrix (tensor) whose elements are given by

$$V_{jk} = \partial z_j/\partial p_k = (\partial\zeta_j/\partial p_k)_{u=\text{const.}} \qquad (4.5)$$

It is symmetric by (3.11). The equation (4.4) is called the Slutsky equation. In interpreting it we note that $\underline{\zeta}d\underline{p}$ is the change in the budget R required to keep the utility constant in compensation for the change $d\underline{p}$ in prices. The second term in (4.4) is called the revenue term (and $dR-\underline{\zeta}d\underline{p}$ is the compensated change in revenue). The first term is the substitution term and can be regarded as giving the change in the distribution of consumption arising from the change in prices but discounting any revenue effect.

By (4.4) and (4.5) we have

$$\partial\zeta_1/\partial p_1 = (\partial\zeta_1/\partial p_1)_{u=\text{const.}} - \zeta_1(\partial\zeta_1/\partial R) . \qquad (4.6)$$

Here the first term on the RHS is ≤ 0 by (3.6). In general, $\partial\zeta_1/\partial R \geq 0$: then $\partial\zeta_1/\partial p_1 \leq 0$ in accordance with intuition. It can however happen that $\partial\zeta_1/\partial R < 0$. If so, the good 1 is called an inferior good (when someone's income increases, he buys less margarine). It is indeed possible for the LHS to be < 0 . If so the good 1 is called a Giffen good: less of it is bought when the price falls. Such goods are more frequent in exam questions than in real life. It is said, however, that when the price of potatoes fell after the Irish famine, then so did the consumption of potatoes.

(The peasants could afford to move to a diet which included items other than potatoes.)

Further reading

Preferences utility and demand. A Minnesota Symposium. (Eds. J.S. Chipman, L. Hurwicz, M.K. Richter and H.F. Sonnenschein.) Harcourt Brace Jovanovich Inc., New York, 1971.

Figure 2.

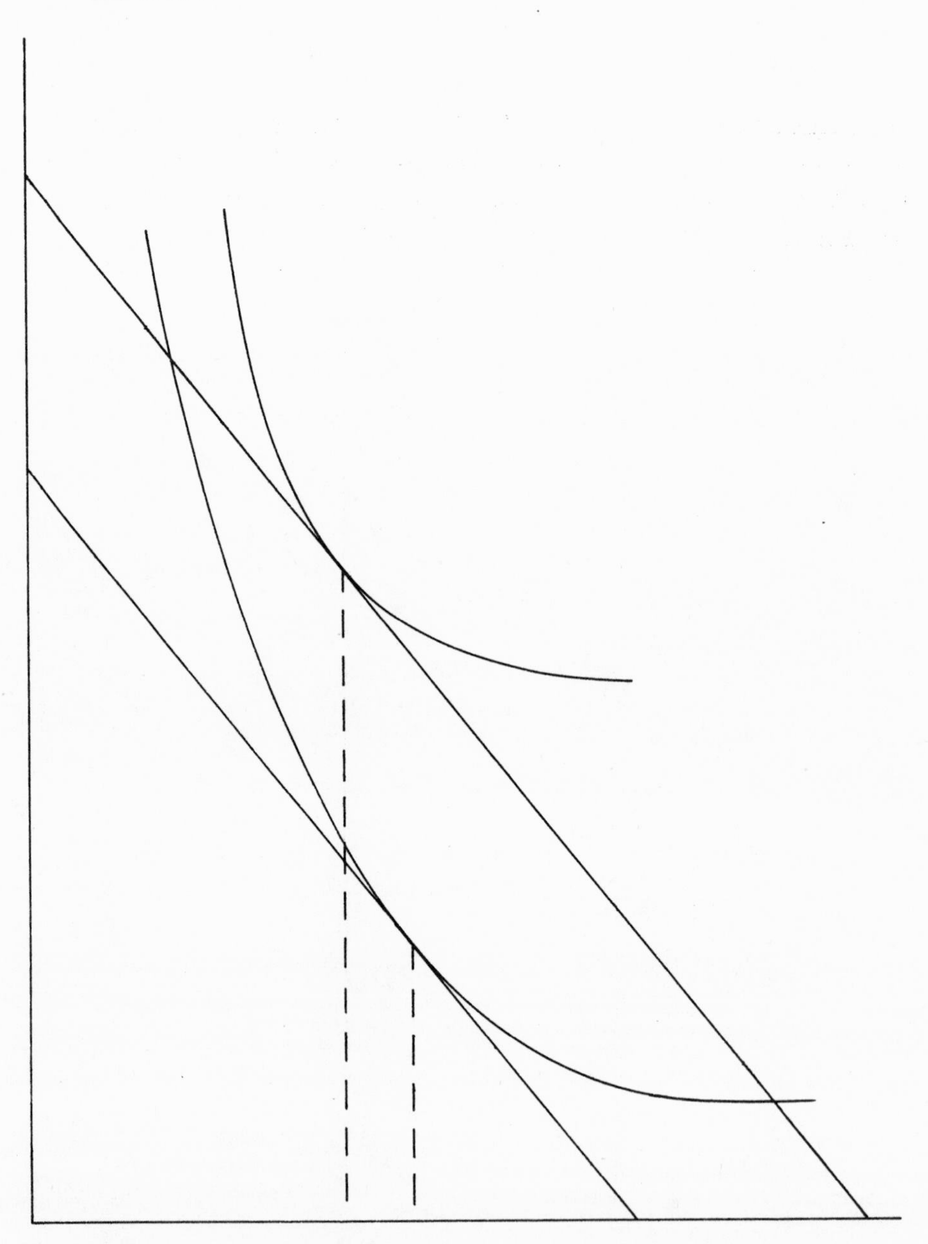

An inferior good

Chapter 1 Exercises

1. Define $\succ$ on the non-negative quadrant of R^2 by $(a,b) \succ (c,d)$ if either $a > c$ or $a = c$ and $b > d$. Show that $\succ$ is an order but is not given by a utility function.

2. Let the order $\succ$ be defined on the non-negative orthant of R^n. Show that the two following statements are equivalent.

   (i) There is a continuous function $u(\underline{x})$ such that $\underline{x} \succ \underline{y}$ precisely when $u(\underline{x}) > u(\underline{y})$.

   (ii) For any $\underline{x}_o$ the sets $\{\underline{x} : \underline{x} \not\succ \underline{x}_o\}$ and $\{\underline{x} : \underline{x}_o \not\succ \underline{x}\}$ are closed.

   [Hint. If (ii) holds, define $u(\underline{x})$ first for the $\underline{x}$ of a countable dense subset, say Q^n.]

3. (Index numbers.) Suppose that there is a transition from a state A to a state B. In State A there is a bundle $\underline{x}^A$ and the price vector is $\underline{p}^A$ and similarly for state B. The Laspeyre index L of state B to base A values the bundles with $\underline{p}^A$ and is by definition

   $$L = \underline{p}^A\underline{x}^B/\underline{p}^A\underline{x}^A .$$

   The Paasche index P values with $\underline{p}^B$ and is by definition

   $$P = \underline{p}^B\underline{x}^B/\underline{p}^B\underline{x}^A .$$

   Now let U be an indifference hypersurface and suppose that $\underline{x}^A$ minimizes $\underline{p}^A\underline{x}$ on U; and similarly for B (so $\underline{x}^A = \underline{z}(\underline{p}^A)$ in the notation of §3). Show that

   $$L \geq 1 \geq P .$$

4. Prove the following "generalized substitution theorem". We suppose that the goods $1,\ldots,r$ are useful only because they contain a certain "generalized good" G. Let the amount of G in a unit of good j be w_j $(1 \leq j \leq r)$; so that

the total amount of good G in a bundle $\underline{x}$ of commodities is $g = \sum_{j \le r} w_j x_j$. Let P be the price of G, so that the price of j is $p_j = w_j P$ $(1 \le j \le r)$. In the notation of §3 let $\underline{z}(\underline{p})$ minimize cost on the indifference hypersurface U. If P increases but p_j is kept fixed for $j > r$, show that the amount of G in $\underline{z}(\underline{p})$ decreases.

5. In the notation of §3, let $d^{(1)}\underline{p}$, $d^{(2)}\underline{p}$ be two infinitesimal changes in the price vector $\underline{p}$ and let $d^{(1)}\underline{z}$, $d^{(2)}\underline{z}$ be the corresponding changes in $\underline{z} = \underline{z}(\underline{p})$. Show that $d^{(1)}\underline{p}d^{(2)}\underline{z} = d^{(2)}\underline{p}d^{(1)}\underline{z}$.

6. An individual chooses the bundle $\underline{x}^{(j)}$ given prices $\underline{p}^{(j)}$ and budget $R^{(j)}$ $(j = 1,2)$. If $\underline{p}^{(1)}\underline{x}^{(1)} > \underline{p}^{(1)}\underline{x}^{(2)}$ show that $\underline{p}^{(2)}\underline{x}^{(1)} > \underline{p}^{(2)}\underline{x}^{(2)}$. If $\underline{p}^{(1)}\underline{x}^{(1)} < \underline{p}^{(1)}\underline{x}^{(2)}$ show that no conclusion can be drawn about the difference between $\underline{p}^{(2)}\underline{x}^{(1)}$ and $\underline{p}^{(2)}\underline{x}^{(2)}$.
[Note. In the first case there is said to be a revealed preference for $\underline{x}^{(1)}$ over $\underline{x}^{(2)}$.]

7. A worker is paid an hourly wage w. He chooses the time he works daily so as to maximize $u(x,y)$, where x is the amount of his daily leisure and y is his daily earnings. Here $u(x,y)$ is a utility function of the usual sort. Construct examples in which an increase in w results in (i) an increase, and (ii) a decrease in the time worked. Is it possible for an increase in w to result in a decrease in the amount y earned?

8. Draw (for $n = 2$) the level curves of a utility function which displays the Giffen goods phenomenon.

9. In the notation of (4.4) show that $\underline{pv} = 1$. If none of the goods is inferior, deduce that $0 \le p_j v_j \le 1$ for each j.

10. (Rationing.) Notation as in §3. The indif-

ference surface U and the price vector $\underline{p}$ are fixed. The good 1 (but no other good) is rationed; that is the quantity x_1 of good 1 chosen must satisfy $x_1 \le c$ for some $c > 0$, where $c < z_1(\underline{p})$. The consumer minimizes $\underline{px}$ on U subject to the condition $x_1 \le c$, say at $\underline{\gamma} = \underline{\gamma}(c)$. Show that $\gamma_1 = c$.

If c varies but $\underline{p}, U$ are kept fixed, show that $d\gamma_2/dc > 0$ or < 0 according as good 2 is a complement of, or a substitute for, good 1.

[*Hint*. Show that

$$\underline{\gamma} = \underline{z}(\lambda, p_2, \ldots, p_n)$$

where $\lambda = \lambda(c) > p_1$ is a decreasing function of c.

Background. D.H. Howard. Rationing, quantity constraints, and consumption theory. *Econometrica* *45* (1977), 399-412.]

11. (Rationing, alternative model.)

(i) Notation as in §4. The prices $\underline{p}$, budget R and utility function are fixed. The good 1 is rationed; that is, the quantity available b (say) satisfies $b < \zeta_1(\underline{p}, R)$. Show that utility is maximized subject to the budget constraint by a bundle of goods $\underline{\beta}$, where $\beta_1 = b$ and

$$\underline{\beta} = \underline{\zeta}(\pi, p_2, \ldots, p_n, R + \pi b - p_1 b)$$

for some $\pi > p_1$.

(ii) Suppose that no good is inferior. If the ration b is allowed to vary, the other parameters being fixed, show that $d\pi/db < 0$.

[*Hint*. Use the Slutsky equation to show that

$$(1 - \pi v_1 + p_1 v_1)db = V_{11} d\pi ,$$

where the coefficient of db is > 0 by exercise 9.]

(iii) Suppose that good 2 is a complement of good 1. Show that $d\beta_2/db > 0$.

12. Fill in the details of the following alternative

model [J.R. Hicks, Revision of demand theory, Chapter 16].

(i) In addition to the n goods labelled $1,\dots,n$ there is a special good "money" labelled O. By definition, the price of money is always unity:

$$p_o = 1 \quad . \qquad (\pounds)$$

We work on a fixed indifference hypersurface $U \subset R^{n+1}$. For any bundle $\underline{q} = (q_1,\dots,q_n)$ of commodities there may or may not be an $m = m(\underline{q}) \geq 0$ such that $(m,\underline{q}) \in U$. We restrict attention to the set S of $\underline{q}$ for which $m(\underline{q})$ exists. There is then a unique price vector $\underline{p} = (p_1,\dots,p_n) = \underline{p}(\underline{q})$ such that the tac-hyperplane to U at $(m,\underline{q})$ is of the shape

$$x_o + \sum_1^n p_j x_j = (1,\underline{p})\underline{x} = \text{const.}$$

Show that

$$\partial p_j/\partial q_k = \partial p_k/\partial q_j \quad . \qquad (\$)$$

(ii) If the common value of (\$) is > 0 we say that goods j,k are quantity-substitutes. For $n = 2$ show that j,k are quantity-substitutes precisely when they are substitutes as defined in the text. Show, further, that this need not be the case when $n > 2$.

(iii) (cf Exercise 3). Let $\underline{q}^A, \underline{q}^B \in S$ and let $\underline{p}^A, \underline{p}^B$ be the corresponding prices. The Laspeyre price-index (weighted for goods) is by definition

$$L^* = \underline{p}^B\underline{q}^A/\underline{p}^A\underline{q}^A$$

and the Paasche price-index is

$$P^* = \underline{p}^B\underline{q}^B/\underline{p}^A\underline{q}^B \quad .$$

Show that

$$L^* \geq P^* \; .$$

CHAPTER 2. PURE EXCHANGE ECONOMY

1. Introduction

We are concerned with bundles of n commodities. There is a finite set E of households h. Each h has a preference relation $\underset{h}{\prec}$ of the kind described in Chapter 1: in particular it is derived from a utility function $u_h(\underline{x})$. The preferences for the different $h \in E$ are supposed to be entirely unrelated. Further, we do not attach any meaning to comparisons between the utility functions of different households h ("there are no interpersonal comparisons of utility"). In a later chapter we shall consider relaxing this restriction (when we get a branch of "welfare economics"), but we shall see that we can get surprisingly far with the present set-up.

We suppose that each household h starts with a bundle of commodities $\underline{w}_h$ (the initial endowment). They exchange these amongst themselves and household h receives $\underline{a}_h$ (the allocation, or final allocation), where

$$\sum_h \underline{a}_h = \sum_h \underline{w}_h . \tag{1.1}$$

Our objective is to consider what allocations are most satisfactory (in senses which have to be made precise) to the $h \in E$ in the light of their preference relations.

We say that an allocation $\underline{a}_h$ is Pareto optimal if there does not exist an allocation $\underline{b}_h$ with

$$\underline{b}_h \underset{h}{\succcurlyeq} \underline{a}_h \qquad \text{(all } h) \tag{1.2}$$

and

$$\underline{b}_h \underset{h}{\succ} \underline{a}_h \qquad \text{(some } h) . \tag{1.3}$$

Clearly any allocation which could be regarded as satisfactory must be Pareto optimal. It is indeed a very minimal condition: for example if $n = 1$ and there are two households with $\underline{w}_1 = \underline{w}_2 = (10)$ then $\underline{a}_1 = (1)$, $\underline{a}_2 = (19)$ is Pareto optimal, although it could hardly be regarded as satisfactory.

There is an alternative definition of Pareto optimality used by some authors (e.g. Arrow-Hahn, who prefer the less loaded term "Pareto efficient"). This is that there should not exist an allocation $\underline{c}_h$ with

$$\underline{c}_h \underset{h}{\succ} \underline{a}_h \qquad \text{(all } h\text{)} . \tag{1.4}$$

Under very weak conditions this is equivalent to the earlier definition. In fact if H is a household for which (1.3) holds, then by continuity there is a $\underline{c}_H < \underline{b}_H$ with $\underline{c}_H \underset{H}{\succ} \underline{a}_H$. The difference $\underline{b}_H - \underline{c}_H$ can then be redistributed among the $h \neq H$ to give an allocation $\underline{c}_h > \underline{b}_h$ $(h \neq H)$.

There is always a large supply of Pareto optimal allocations. Let $\lambda_h > 0$ $(h \in E)$ be arbitrary. The continuous function

$$\Sigma\, \lambda_h u_h(\underline{a}_h) \tag{1.5}$$

attains its maximum in the compact set (1.1). Any $\underline{a}_h$ which gives the maximum is clearly Pareto optimal.

A more restrictive condition on allocations was (essentially) introduced by Edgeworth, namely the <u>core</u>. Let $\{\underline{a}_h\}$ be an allocation. We say that a subset $S \subset E$ is a <u>blocking coalition</u> for $\{\underline{a}_h\}$ if there exists $\underline{b}_s$ $(s \in S)$ such that

$$\sum_S \underline{b}_s = \sum_S \underline{w}_s , \tag{1.6}$$

$$\underline{b}_s \underset{s}{\succsim} \underline{a}_s \qquad \text{(all } s \in S\text{)} \tag{1.7}$$

$$\underline{b}_s \underset{s}{\succ} \underline{a}_s \qquad \text{(some } s \in S\text{)} . \tag{1.8}$$

We say that $\{\underline{a}_h\}$ is in the core if there are no blocking coalitions. An allocation in the core is Pareto optimal (take S = E). By considering coalitions consisting of a single household we see that core allocations satisfy

$$\underline{a}_h \underset{h}{\succsim} \underline{w}_h \qquad \text{(all } h \in E) . \qquad (1.9)$$

There is a third notion, due to Walras. We say that the allocation $\{\underline{a}_h\}$ is competitive if there is a price vector $\underline{p} > \underline{0}$ such that for each h the bundle $\underline{a}_h$ maximizes $u(\underline{x})$ subject to the budget restraint

$$\underline{px} \leq \underline{pw}_h \ . \qquad (1.10)$$

Then

$$\underline{pa}_h = \underline{pw}_h \ . \qquad (1.11)$$

The wording has been chosen so that it makes sense without the hypothesis that indifference hypersurfaces are convex. If we do suppose that indifference surfaces are strictly convex, then $\underline{a}_h$ is uniquely determined by $\underline{p}$ and the $\underline{w}_h$ (and the functions $u_h(\underline{x})$). Of course if we choose $\underline{p}$ at random and let each h maximize its utility subject to (1.10), then there is no reason to expect that we shall get an allocation: the condition (1.1) need not be satisfied.

Lemma 1.1. A competitive allocation is in the core.

Proof. Let the competitive allocation be $\{\underline{a}_h\}$ corresponding to price vector $\underline{p}$. Suppose, if possible, that a blocking coalition S and bundles $\underline{b}_h$ exist satisfying (1.7), (1.8).

Since $\underline{a}_h$ maximizes utility subject to (1.10), we have

$$\underline{pb}_s \geq \underline{pw}_s \qquad \text{(all } s \in S) , \qquad (1.12)$$

$$\underline{pb}_s > \underline{pw}_s \qquad \text{(some } s \in S) . \qquad (1.13)$$

But (1.6) implies

$$\sum_S \underline{pb}_s = \sum_S \underline{pw}_s \ . \qquad (1.14)$$

Contradiction!

In the rest of this chapter we shall prove:

There is always a competitive equilibrium. (§3).

If the condition of convexity on indifference hypersurfaces is omitted, it is possible for the core to be empty (and so, a fortiori, there are no competitive allocations). (§5).

In general not every allocation in the core is competitive. (§2).

The following notion plays an important rôle. Let E be an economy as described above, with households h each having its preference relation $\underset{h}{\prec}$ and initial endowment $\underline{w}_h$. Let $N > 1$ be an integer. For each $h \in E$ the replicated economy NE contains N households $\chi(1,h),\ldots,\chi(N,h)$ of type h, each with preference relation $\underset{h}{\prec}$ and initial endowment $\underline{w}_h$. If $\{\underline{a}_h\}$ is an allocation for E, then the allocation of NE which gives $\underline{a}_h$ to $\chi(j,h)$ $(1 \le j \le N)$ $(h \in H)$ is said to be the replication of $\{\underline{a}_h\}$. In §4 we show that $\{\underline{a}_h\}$ is competitive precisely when its replication is in the core of NE for every N.

It is convenient to prove here

Lemma 1.2. Suppose that all indifference hypersurfaces are strictly convex. Then every core allocation of NE is the replication of some core allocation of E.

Proof. Consider the allocation of NE which gives $\underline{a}(j,h)$ to household $\chi(j,h)$. Without loss of generality

$$\underline{a}(1,h) \underset{h}{\precsim} \underline{a}(2,h) \ldots \underset{h}{\precsim} \underline{a}(N,h) . \tag{1.15}$$

Put

$$\underline{b}(h) = N^{-1} \sum_j \underline{a}(j,h) , \tag{1.16}$$

so $\{\underline{b}(h)\}$ is an allocation for E. It is easy to see (using the strict convexity hypothesis) that

$$\underline{a}(1,h) \underset{h}{\prec} \underline{b}(h) , \tag{1.17}$$

unless all the $\underline{a}(j,h)$ $(1 \le j \le N)$ are equal. Thus the coalition of the $\chi(1,h)$ $(h \in H)$ is blocking unless the allocation $\{\underline{a}(j,h)\}$ is the replication of $\{\underline{b}(h)\}$. Hence if $\{\underline{a}(j,h)\}$ is in the core it is a replication of $\{\underline{b}(h)\}$, which must clearly be in the core of E.

2. The Edgeworth box

In this section, following Edgeworth, we consider the case when there are two goods ($n = 2$) and two households, which we shall label A and B. We shall also suppose, though it is not vital to the argument, that each starts with an initial endowment of just one of the goods, say

$$\underline{w}_A = (0,1) \; ; \; \underline{w}_B = (1,0) \; . \tag{2.1}$$

We restrict attention to the case when the indifference curves (= indifference hypersurfaces for $n = 2$) are strictly convex.

If the allocation of A is

$$\underline{a}_A = (x,y) \; , \tag{2.2}$$

then the allocation of B is

$$\underline{a}_B = (1-x, \; 1-y) \; . \tag{2.3}$$

We may therefore represent both $\underline{a}_A$ and $\underline{a}_B$ by the single point (x,y) in the unit square

$$0 \le x \le 1 \; , \; 0 \le y \le 1 \tag{2.4}$$

(the <u>Edgeworth box</u>). The allocation $\underline{a}_A$ is measured from the origin in the standard way while $\underline{a}_B$ is measured backward from $(1,1)$. Both $\underline{w}_A$ and $\underline{w}_B$ are thus represented by $(0,1)$.

The indifference curves for A are convex with respect to the origin. The indifference curves for B in the box representation are convex with respect to $(1,1)$ (and so concave to the origin).

It is easy to see that (x,y) represents a Pareto optimal allocation precisely when the indifference curves for A and B touch at (x,y). They therefore

Figure 3.

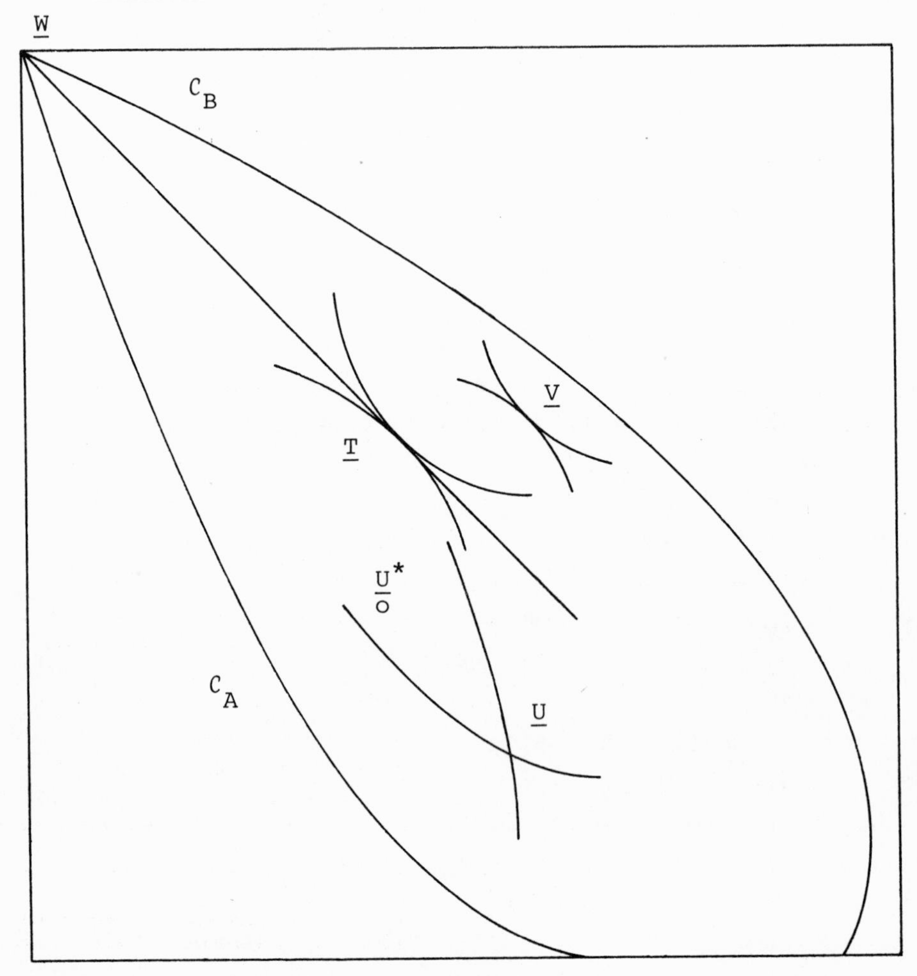

W gives the initial allocation. C_A is A's indifference curve through W. He will not agree to anything below C_A because he would prefer the status quo. Similarly for C_B. The two parties will not agree to U, where the indifference curves cross, since at e.g. U^* they would both be better off. The contract curve consists of points such as V where the indifference curves touch. A point such as T where the common tangent passes through W is a competitive equilibrium.

lie on a curve. If (x,y) represents a point of the core, we require in addition that $\underline{a}_A \succsim_A \underline{w}_A ; \underline{a}_B \succsim_B \underline{w}_B$: that is that (x,y) lies between the A- and B-indifference curves through $(0,1)$. The core is thus represented by a curvilinear arc, which Edgeworth calls the contract curve.

We now consider the condition that (x,y) represent a competitive allocation. It is easy to see that for any price-vector $\underline{p}$ the budget restraints

$$\underline{pa}_A = \underline{pw}_A ; \underline{pa}_B = \underline{pw}_B \tag{2.5}$$

are represented by the same straight line L in the box through $(0,1)$. The point (x,y) maximizes A-utility on L if L is tangential to the A-indifference curve through (x,y) ; and similarly for B . Hence (x,y) is a point on the contract curve at which the common tangent to the A- and B- indifference curves passes through $(0,1)$. It is left to the reader to show that such a point always exists: we shall prove the corresponding fact for arbitrary numbers of households and goods later.

Now, following Edgeworth, we use his box to investigate the replicated economy NE , where $N > 1$ is an integer and E is the economy described above. By Lemma 1.2 all individuals of the same type receive the same allocation. We may thus represent a point of the core of NE by a point (x,y) in the Edgeworth box. It clearly must be in the core of E , but there are now further restrictions. In fact for any $r \leq N$, $s \leq N$, a set S of r individuals of type A and s individuals of type B may constitute themselves a coalition and allocate themselves $\underline{b}_A, \underline{b}_B$, where

$$r\underline{b}_A + s\underline{b}_B = r\underline{w}_A + s\underline{w}_B \; . \tag{2.6}$$

Here $\underline{b}_A, \underline{b}_B$ are not, in general, represented by the same point in the Edgeworth box, but since

$$r(\underline{w}_A - \underline{b}_A) + s(\underline{w}_B - \underline{b}_B) = 0 \ , \tag{2.7}$$

the two representative points $\underline{x}_A$, $\underline{x}_B$ (say) lie on the same line through $(0,1)$. Further, any two points $\underline{X}_A$, $\underline{X}_B$ collinear with $(0,1)$ can be approximated arbitrary closely by $\underline{x}_A$, $\underline{x}_B$ for suitable choice of r,s provided that N is large enough. It follows that the core of NE shrinks to the competitive allocations as N increases. [There may be more than one, see Exercise 2.] We leave the detail to the reader, as we shall later deal with the case of arbitrarily many households and commodities.

3. Existence of competitive allocations

In this section we prove the existence of competitive allocations. The argument makes essential use of the assumption that indifference hypersurfaces are strictly convex.

We retain the notation of §1. Since prices occur homogeneously we can without loss of generality restrict attention to price vectors $\underline{p}$ for which

$$\Sigma \, p_j = 1 \quad , \tag{3.1}$$

(normalized price vectors). For every normalized $\underline{p}$ satisfying

$$\underline{p} \gg \underline{0} \tag{3.2}$$

and for household h there is a unique bundle $\underline{c}_h(\underline{p})$ which maximizes h-utility subject to the budget restraint

$$\underline{px} = \underline{pw}_h \ . \tag{3.3}$$

Here $\underline{w}_h$ is the initial endowment. The vector $\underline{c}_h(\underline{p})$ is clearly a continuous function of $\underline{p}$. If

$$\sum_h \underline{c}_h(\underline{p}) = \sum_h \underline{w}_h \ , \tag{3.4}$$

then $\underline{c}_h(\underline{p})$ is a competitive allocation. This leads to the consideration of the excess demand function

$$\underline{e}(\underline{p}) = \sum_h \underline{c}_h(\underline{p}) - \sum_h \underline{w}_h \ . \tag{3.5}$$

Then

$$\underline{pe}(\underline{p}) = \underline{0} \tag{3.6}$$

by (3.3), and we wish to find a $\underline{p}$ with

$$\underline{e}(\underline{p}) = \underline{0} \; . \tag{3.7}$$

We must, however, also consider price vectors $\underline{p}$ which do not satisfy (3.2), i.e. for which some goods are free, and then $\underline{c}_h(\underline{p})$ may not be well-defined. We must therefore use a technical trick. Let $t > 0$ be real and put

$$\underline{t} = (t,\ldots,t) \quad . \tag{3.8}$$

We suppose t is so large that

$$\sum_h \underline{w}_h << \underline{t} \tag{3.9}$$

but otherwise arbitrary.

Now for any $\underline{p}$ (not necessarily satisfying (3.2)) the continuous function $u_h(\underline{x})$ attains its maximum on the set defined by (3.2) and

$$\underline{x} \le \underline{t} \; . \tag{3.10}$$

By the strict convexity of the indifference hypersurfaces, the maximum is attained at precisely one point, which we denote by $\underline{c}_h^*(\underline{p})$. It is not difficult to see that $\underline{c}_h^*(\underline{p})$ depends continuously on $\underline{p}$. Hence

$$\underline{e}^*(\underline{p}) = \sum_h c_h^*(\underline{p}) - \sum \underline{w}_h \tag{3.11}$$

depends continuously on $\underline{p}$ and satisfies

$$\underline{pe}^*(\underline{p}) = 0 \; . \tag{3.12}$$

We are now in a position to apply the following

<u>Lemma 3.1</u>. Let $\underline{v}(\underline{p})$ be an n-dimensional vector depending continuously on $\underline{p}$ in the set

$$\sum p_j = 1 \qquad p_j \ge 0 \; . \tag{3.13}$$

Suppose that

$$\underline{pv}(\underline{p}) = 0 \tag{3.14}$$

for all $\underline{p}$. Then there is a $\underline{p}^*$ such that

$$\underline{v}(\underline{p}^*) \le \underline{0} \; . \tag{3.15}$$

Before proving the lemma, we show that it gives the existence of a competitive allocation on putting $\underline{v}(\underline{p}) = \underline{e}^*(\underline{p})$. The hypotheses are satisfied by (3.12), and so there is a $\underline{p}^*$ with

$$\underline{e}^*(\underline{p}^*) \leq \underline{0} \;; \tag{3.16}$$

that is

$$\sum_h \underline{c}_h^*(\underline{p}^*) \leq \sum_h \underline{w}_h \;. \tag{3.17}$$

By (3.9) and (3.17) we have

$$\underline{c}_h^*(\underline{p}^*) \ll \underline{t} \qquad \text{(all } h \in E) \;. \tag{3.18}$$

On recalling the definition of the $\underline{c}_h^*(\underline{p})$ and the fact that $u_h(\underline{x})$ increases with $\underline{x}$ we deduce that there is actually equality in (3.17). Further, (3.18) shows that $\underline{c}_h^*(\underline{p}^*)$ gives the maximum of $u_h(\underline{x})$ subject only to (3.3) (with $\underline{p} = \underline{p}^*$), that is $\underline{c}_h^*(\underline{p}^*) = \underline{c}_h(\underline{p}^*)$. Hence $\underline{p}^*$ gives a competitive allocation, as required.

It remains only to prove Lemma 3.1. We do this by constructing a continuous map of the simplex (3.13) into itself and applying Brouwer's fixed point theorem (see Appendix B). More precisely, we define $T(\underline{p}) = \underline{q}$ by

$$q_j = (p_j + \delta_j)/Q \tag{3.19}$$

where

$$Q = \sum_k (p_k + \delta_k) = 1 + \sum_k \delta_k \tag{3.20}$$

and

$$\delta_j = \max(v_j, 0) \;,\; \underline{v} = \underline{v}(\underline{p}) \;. \tag{3.21}$$

(This has the obvious economic interpretation: if the good j is in excess demand, increase its price.) By Brouwer's theorem there is a $\underline{p}^*$ with $T(\underline{p}^*) = \underline{p}^*$. Then

$$p_j^*(1 + \sum_k \delta_k) = p_j^* + \delta_j \;, \tag{3.22}$$

$$\delta_j = p_j^* \sum \delta_k \;. \tag{3.23}$$

If $\sum \delta_k = 0$, then $\delta_k = 0$ for all k and we are done. If $p_j^* = 0$, then $\delta_j = 0$. Further,

$$p_j^* > 0 \Rightarrow \delta_j > 0 \Rightarrow v_j^* > 0 \;; \tag{3.24}$$

and so

$$\sum p_j^* v_j^* > 0 \;, \tag{3.25}$$

in contradiction to (3.14).

Addendum

The condition that the indifference hypersurfaces are strictly convex can be relaxed to mere convexity by the use of Kakutani's generalization of Brouwer's fixed point theory [see Addendum to Appendix B]. The set of $\underline{x}$ which maximize h-utility subject to (3.3) is now, in general, not a single bundle $\underline{c}_h(\underline{p})$ but a closed convex set. It follows that the excess demand function $\underline{e}(\underline{p})$ must be replaced by a closed convex set $E(\underline{p})$ depending on $\underline{p}$. On the other hand, Kakutani's theorem gives a generalization of Lemma 3.1 in which $\underline{v}(\underline{p})$ is replaced by a set $V(\underline{p})$ such that $\underline{pv} = 0$ for all $\underline{v} \in V(\underline{p})$ and depending upper semi-continuously on $\underline{p}$; the conclusion being that there are $\underline{p}^*$ and $\underline{v}^* \in V(\underline{p}^*)$ such that $\underline{v}^* \leq \underline{0}$.

4. Replicated economies

The main result of this section does not require any assumption about the convexity of indifference hypersurfaces. This increased generality can be obtained at the cost of very little additional complication in the proof.

Theorem 4.1. Let $\{\underline{a}_h\}$ be an allocation of E whose replication is in the core of NE for all $N > 1$. Then $\{\underline{a}_h\}$ is competitive.

Note. Convexity of the indifference hypersurfaces is not assumed, so "competitive" must be understood in the more general sense discussed in §1.

Proof. For each $h \in E$ let

$$\Delta_h = \{\underline{x} : \underline{x} \underset{h}{\succcurlyeq} \underline{a}_h\} \tag{4.1}$$

and

$$\Gamma_h = \Delta_h - \underline{w}_h = \{\underline{x} : \underline{x} + \underline{w}_h \in \Delta_h\} . \tag{4.2}$$

Let Δ_h^o, Γ_h^o be the interiors of Δ_h, Γ_h: so

$$\underline{x} \in \Delta_h^o \Rightarrow \underline{x} \underset{h}{\succ} \underline{a}_h \, . \tag{4.2 bis}$$

We denote by C the convex cover of $\underset{h}{\cup} \Gamma_h^o$, so C is open (cf. Appendix A).

Suppose first that

$$\underline{O} \in C \, . \tag{4.3}$$

Then

$$\underline{O} = \sum_{1 \leq k \leq K} \lambda_k \underline{y}_k \qquad (\lambda_k > 0, \ \Sigma \ \lambda_k = 1) \tag{4.4}$$

for some $K \geq 1$ and for some

$$\underline{y}_k \in \Gamma_{h(k)}^o \tag{4.5}$$

with

$$h(k) \in E \qquad (1 \leq k \leq K). \tag{4.6}$$

Since the Γ_h^o are open, (4.4) remains valid if we replace the λ_k by rational numbers ℓ_k close enough to them, and at the same time replace the $\underline{y}_k$ by appropriate $\underline{z}_k \in \Gamma_{h(k)}^o$ close enough to them. On multiplying up by the common denominator of the ℓ_k we thus have

$$\underline{O} = \sum_{1 \leq k \leq K} L_k \underline{z}_k \, , \tag{4.7}$$

where

$$\underline{z}_k \in \Gamma_{h(k)}^o \, , \tag{4.8}$$

and the

$$L_k > 0 \tag{4.9}$$

are integers.

Let $N > \Sigma \ L_k$ and consider the replicated economy NE. Let S be a coalition which is the disjoint union of the sets $S(k)$ consisting of L_k households of type $h(k)$ for $1 \leq k \leq K$. To each of the households in $S(k)$ we make the allocation

$$\underline{b}_k = \underline{w}_k + \underline{z}_k \, . \tag{4.10}$$

This is a re-allocation of the initial endowments of S, by (4.7). Further,

$$\underline{b}_k \in \Delta_{h(k)}^o \tag{4.11}$$

by (4.2); so

$$\underline{b}_k \underset{h(k)}{\succ} \underline{a}_{h(k)} \tag{4.12}$$

by (4.2 bis). Hence S is a blocking coalition, contrary to the assumption that the replication of $\{\underline{a}_n\}$ is in the core.

Thus (4.3) leads to a contradiction, and we must have

$$\underline{0} \notin C . \tag{4.13}$$

Since C is convex, there is by Theorem 1 of Appendix A a $\underline{q} \neq \underline{0}$ such that

$$\underline{qc} > 0 \qquad \text{(all } \underline{c} \in C) . \tag{4.14}$$

We now show that $\underline{q}$ is a price vector. Let $\underline{e}_1 = (1,0,\ldots,0)$. The basic properties of $\prec$ imply that $\underline{x} + t\underline{e}_1 \in \Delta_h^o$ whenever $\underline{x} \in \Delta_h^o$ and $t \geq 0$. Hence $\underline{c} + t\underline{e}_1 \in C$ whenever $\underline{c} \in C$. On replacing $\underline{c}$ by $\underline{c} + t\underline{e}_1$ in (4.14) and making $t \to \infty$ we see that $q_1 \geq 0$. Similarly $q_j \geq 0$ $(2 \leq j \leq n)$; i.e. $\underline{q}$ is a price vector.

Since $\Gamma_h^o \subset C$ and Γ_h is the closure of Γ_h^o (by (1.5) of Chapter 1), it follows from (4.14) that

$$\underline{qc} \geq 0 \qquad \text{(all } \underline{c} \in \Gamma_h) . \tag{4.15}$$

By (4.1), (4.2) we have

$$\underline{x} \underset{h}{\succsim} \underline{a}_h \Rightarrow \underline{qx} \geq \underline{qw}_h . \tag{4.16}$$

In particular,

$$\underline{qa}_h \geq \underline{qw}_h \qquad (h \in E) . \tag{4.17}$$

But $\Sigma\, \underline{a}_h = \Sigma\, \underline{w}_h$, and so

$$\underline{qa}_h = \underline{qw}_h . \tag{4.18}$$

Hence the allocation $\{\underline{a}_h\}$ is competitive, corresponding to the price-vector $\underline{q}$.

Comment. Aumann has introduced a generalized economy in which the set of households is a measure space without atoms (cf also book of Hildenbrand). No assumption of convexity is required on the preference relations but they depend continuously, in an appropriate sense, on h. It is shown that the core consists precisely of the competitive allocations.

5. Non-convex preferences

In this section we sketch an example to show that the core can be empty if the condition on convexity of indifference hypersurfaces is omitted.

The number of goods is $n = 2$. All households have the same utility function

$$u(x,y) = x^2 + y^2 \tag{5.1}$$

and the same initial endowment

$$\underline{w} = (1,1) . \tag{5.2}$$

Let N be the number of households in E.

When $N = 2$ it is easy to see that a core allocation is (2,0), (0,2). Note however that (in contrast to the convex case) there is no core allocation in which the final allocations are the same; that is the two households have here to agree which is to get all of good 1 and which all of good 2.

We now indicate briefly steps to show that for $N = 3$ there is no core allocation. Suppose, if possible, that

$$\underline{a}_j = (x_j, y_j) \qquad (1 \le j \le 3) \tag{5.3}$$

is in the core.

(i) Suppose that $\underline{a}_j >> \underline{0}$ $(j = 1,2)$. Then there is a $\underline{d}$ such that

$$\underline{a}_1 + \underline{d} \succ \underline{a}_1 \ , \ \underline{a}_2 - \underline{d} \succ \underline{a}_2 .$$

Hence at least two of $\underline{a}_1, \underline{a}_2, \underline{a}_3$ are on the co-ordinate axes.

(ii) On considering coalitions of a single member we have

$$x_j^2 + y_j^2 \ge 2 \qquad (1 \le j \le 2) .$$

(iii) On considering coalitions with $N = 2$, there can be at most one j with $x_j^2 + y_j^2 < 4$.

(iv) Suppose that $\underline{a}_1 = (a,0)$, $\underline{a}_2 = (b,0)$ with $a \le b$. Then $a \ge 2^{\frac{1}{2}}$, $b \ge 2$, so $a+b > 3$, a contradiction.

(v) Suppose $\underline{a}_1 = (a,0)$, $\underline{a}_2 = (0,b)$, so $\underline{a}_3 = (3-a,3-b)$ with $a \le b$. Then $a \ge 2^{\frac{1}{2}}$, $b \ge 2$ and

$$(3-a)^2 + (3-b)^2 < 4 .$$

Hence $a \ge 2$ and $\underline{a}_1 = (2,0)$, $\underline{a}_2 = (0,2)$, $\underline{a}_3 = (1,1)$. But this is not in the core since $\underline{a}_1, \underline{a}_3$ are blocking coalition. (Indeed $\underline{b}_1 = (\varepsilon,2) \succ \underline{a}_1$; $\underline{b}_3 = (2-\varepsilon,0) \succ \underline{a}_3$ if $\varepsilon > 0$ is small.)

Further reading

K. Arrow and F.H. Hahn. Competitive analysis. Holden-Day; Oliver and Boyd. 1971.

F.Y. Edgeworth. Mathematical Psychics. Kegan Paul, London. 1881. Reprinted as: Series of reprints of scarce tracts in economics and political science. No.10. London School of Economics, 1932.

W. Hildenbrand. Core and equilibria of a large economy. Princeton Univ. Press, 1974.

Chapter 2 Exercises

1. Two traders deal in goods labelled 1,2. The first has initial endowment (0,1) and utility function $x^2 + 3xy + 2y^2$: the second has initial endowment (1,0) and utility function xy. Find the contract curve and the point of competitive equilibrium.
2. Show that there may be several distinct competitive equilibria, even when the indifference curves are strictly convex.
   [Hint. Use the Edgeworth box and consider the condition that a point on the contract curve is a competitive equilibrium.]
3. Generalize the theory of the Edgeworth box to the case where the two agents A,B have initial endowments $\underline{w}_A, \underline{w}_B$ which may contain both goods. If $\underline{w}_A + \underline{w}_B = (1,1)$, show that all the contract curves are portions of a single curve (the extended contract curve).
4. For a given utility function $u(x,y)$ an Engel curve (NOT the red angel) is the locus of the bundles (ξ,η) chosen for the budget $px+qy = B$, where the prices p,q are fixed but B varies. Defining the marginal propensity to consume good 1 as $d\xi/dB$, show that the slope of the Engel curve is the ratio of the two marginal propensities to consume. If $u(x,y)$ is continuously differentiable and has strictly convex indifference curves, show that there is precisely one Engel curve through each point $(x,y) >> (0,0)$.
5. In an Edgeworth box consider the Engel curves $E_A(p_1,p_2)$, $E_B(p_1,p_2)$ of the two agents A,B for the pair of prices (p_1,p_2). Show that they intersect (if at all) on the extended contract

curve C, and that C lies between them.

6. Country A produces only manufactured goods and country B only food. The rates of production are constant. They trade only with each other and achieve a competitive equilibrium with respect to utility functions $u_A(x,y)$, $u_B(x,y)$ where x,y are the rates of consumption of manufactured goods and food respectively. Let p_M, p_F be the equilibrium prices of manufactures and food (in terms of some trading currency), and define p_M/p_F to be A's _terms of trade_.

   Now suppose that A gives B a fixed (small) annual amount of manufactures as aid, but that otherwise they trade as before. Show that A's terms of trade will deteriorate (i.e. diminish) precisely when

$$\frac{C(A,M)}{C(A,F)} > \frac{C(B,M)}{C(B,F)} ,$$

   where $C(A,F)$ is A's marginal propensity to consume food (at the equilibrium quantities and prices), etc.
   [_Hint_. Consider the Engel curves through the equilibrium point. Cf. P.A. Samuelson, _Economic J._ 62 (1952), 278-304 = _Collected Papers_ II, 985-1011.]

7. Let $\{\underline{a}_h\}$ be a Pareto optimal allocation for the economy E with convex indifference hypersurfaces. Show that there are prices $\underline{p}$ such that $\underline{a}_h$ maximized h-utility subject to the budget constraint

$$\underline{px} \le \underline{pa}_h .$$

   To what does this correspond for the Edgeworth Box?
   [_Note_ It is not, of course, claimed that $\underline{pa}_h = \underline{pw}_h$.

   Hint. In the notation of §4 show that $\underline{O}$ is

is not in the convex cover of the union of the $\Delta_h^o - \underline{a}_n$.]

8. (Arrow-Debreu economy. Bowdlerised version.) In this economy, in addition to the finite set E of households h (with the properties described in §1) there is a finite set J of <u>producers</u> j .

Each producer j may take a bundle of commodities $\underline{x}$ (the <u>input</u>) and transform it into a bundle $\underline{z}$ (the <u>output</u>) depending on $\underline{x}$ and j . The set Y_j of $\underline{y} = \underline{z} - \underline{x}$ (the <u>net product</u>) is j's <u>production set</u>. We suppose that Y_j is strictly convex, closed and bounded, and that $\underline{O} \in Y_j$.

Let $\underline{w}_h$ $(h \in E)$ be the initial allocations of the households, and suppose that producer j chooses net product $\underline{y}_j \in Y_j$ $(j \in J)$. Then an allocation $\underline{b}_h$ to households is <u>feasible</u> if

$$\sum_h \underline{b}_h = \sum_h \underline{w}_h + \sum_j \underline{y}_j \ .$$

We suppose, further, that the producers are owned by the households. Household h owns a share θ_{hj} of producer j , where $\theta_{hj} \geq 0$ and $\sum_h \theta_{hj} = 1$ (all $j \in J$) .

Under prices $\underline{p}$, producer j chooses $\underline{y}_j \in Y_j$ so as to maximize his profit $\underline{py}_j$. Then household h chooses bundle $\underline{x}_h$ to maximize its utility subject to the budget constraint

$$\underline{px}_h \leq \underline{pw}_h + \sum_j \theta_{hj} \underline{py}_j \ .$$

Show that $\underline{p}$ can be chosen so that the allocation $\underline{x}_h$ is feasible.

[<u>Hint</u>. §3. <u>References</u> K.J. Arrow and G.Debreu, Existence of an equilibrium for a competitive economy. <u>Econometrica</u> <u>22</u> (1954), 265-290 : G. Debreu, <u>Theory of value</u> (Yale University Press, 1959), §5.7 : Arrow-Hahn loc.cit.]

9. Fill in the steps of the following argument which shows that, when the number of households is large compared with the number of commodities, every core allocation is in some sense nearly competitive. We need not suppose that the indifference hypersurfaces are convex. Let $\underline{w}_h$ be the initial endowment and $\underline{a}_h$ a core allocation. Denote by Γ_h^o the interior of the set defined by (4.2), and put

$$\Lambda_h = \Gamma_h^o \cup \{\underline{O}\} .$$

(i) $\Sigma\, \Lambda_h$ cannot contain an $\underline{x} < \underline{O}$. [For otherwise there is a coalition $S \subset E$ and $\underline{b}_s \in \Gamma_s^o$ $(s \in S)$ such that $\Sigma\, \underline{b}_s < \underline{O}$. Show that S blocks $\{\underline{a}_h\}$.]

Now let $\underline{k}$ be any bundle such that

$$\underline{w}_h \le \underline{k} \quad (\text{all } h \in E) .$$

(ii) The convex cover $\text{con}(\Sigma\, \Lambda_h)$ cannot contain an $\underline{x} < -n\underline{k}$, where n is the number of commodities. [For otherwise by the Shapley-Folkman theorem (Appendix A, Exercise 2) there are $\underline{\ell}_h \in \text{con}(\Lambda_h)$ with $\Sigma\, \underline{\ell}_h < -n\underline{k}$ and $\underline{\ell}_h \notin \Lambda_h$ for at most n households h. Replace these exceptional $\underline{\ell}_h$ by $\underline{O}$ and apply (i).]

(iii) There is a price vector $\underline{p}$ with $\Sigma\, p_j = 1$ such that

$$\underline{x} \in \Sigma\, \Lambda_h \Rightarrow \underline{px} \ge -nK ,$$

where $K = \max k_j$. [Apply the separation Theorem 2 of Appendix A to (ii).]

(iv) $\sum_h |\underline{pa}_j - \underline{pw}_h| \le 2nK .$

[Let S be the set of h with $\underline{pa}_h < \underline{pw}_h$. Then $\sum_S (\underline{a}_s - \underline{w}_s)$ is in the closure of $\Sigma\, \Lambda_h$. But $\Sigma\, \underline{a}_h = \Sigma\, \underline{w}_h$.]

(v) Let

$$\mu_h = \inf(\underline{px} - \underline{pw}_h) \quad (x \succ_h \underline{a}_h) .$$

Then

$$\sum_h |\mu_h| \leq 2nK .$$

[For $\mu_h \leq \underline{p}(\underline{a}_h - \underline{w}_h)$. Let S be the set of h with $\mu_h < 0$. Argue as in (iv).]

[Comment. The point is that the estimates are independent of the number of households, provided only that $\underline{w}_h \leq \underline{k}$.
Reference. R.M. Anderson. An elementary core equivalence theorem. Econometrica 46 (1978), 1483-1487.]

10. Deduce Theorem 4.1 from the preceding exercise.
11. Hypotheses and notation as in Exercise 9, except that the allocation $\{\underline{a}_h\}$ is not necessarily in the core. Suppose, however, that it is not blocked by any coalition of $\leq M$ members, where M is some number less than the number N of households. Show that the conclusions of (iv),(v) continue to hold if $2nK$ is replaced by $2nHK$, where H is the least integer such that $MH \geq N$. [Hint. Partition the households into M sets I_m $(1 \leq m \leq M)$, each with at most H members. Put $\Theta_m = \cup \Lambda_h$ $(h \in I_m)$ and show that (iii) holds with $\Sigma\, \Theta_m$ instead of $\Sigma\, \Lambda_h$. Let T be any set of $\leq M$ households no two of which are in the same I_m . Show that (iv), (v) hold (with the same bound $2nK$) if Σ is replaced by the sum over the $h \in T$. But E^h is the union of at most H such T . Reference. A. Mas-Colell. A refinement of the core equivalence theorem. Economics Letters 3 (1979), 307-310.]

CHAPTER 3. THEORY OF THE FIRM

1. Introduction

In this chapter we make a further stride towards realism and consider a situation in which there are firms which can create certain goods (outputs) by the use of other goods (inputs).

We operate in a monetary economy of an especially simple type. There is a single good (money) which can be freely exchanged. It can be neither created nor destroyed, and it does not enter into any of the manufacturing processes considered. All prices are in terms of money. We start by considering only the monetary cost of the output, assuming that the prices of the inputs are constant.

2. Supply and demand

We now consider a single good which is traded between a number of producers and consumers. We suppose that all transactions take place at the same price. At any price p there is a certain demand $d(p)$ from the consumers. We naturally suppose that $d(p)$ decreases as p increases and, for simplicity, that $d(p)$ is continuous. Similarly, at any price p there is the supply $s(p)$, the amount that the producers will produce for price p. Here $s(p)$ increases with p and is supposed continuous. Under reasonable conditions there is then a unique price P which equates supply and demand:

$$d(P) = s(P) = Q \quad \text{(say)} . \tag{2.1}$$

Economic forces are supposed to lead to a situation at

which the price is P and a quantity Q is produced and consumed. In what follows we suppose $s(p), d(p)$ are graphed with p on the y-axis.

Figure 4.

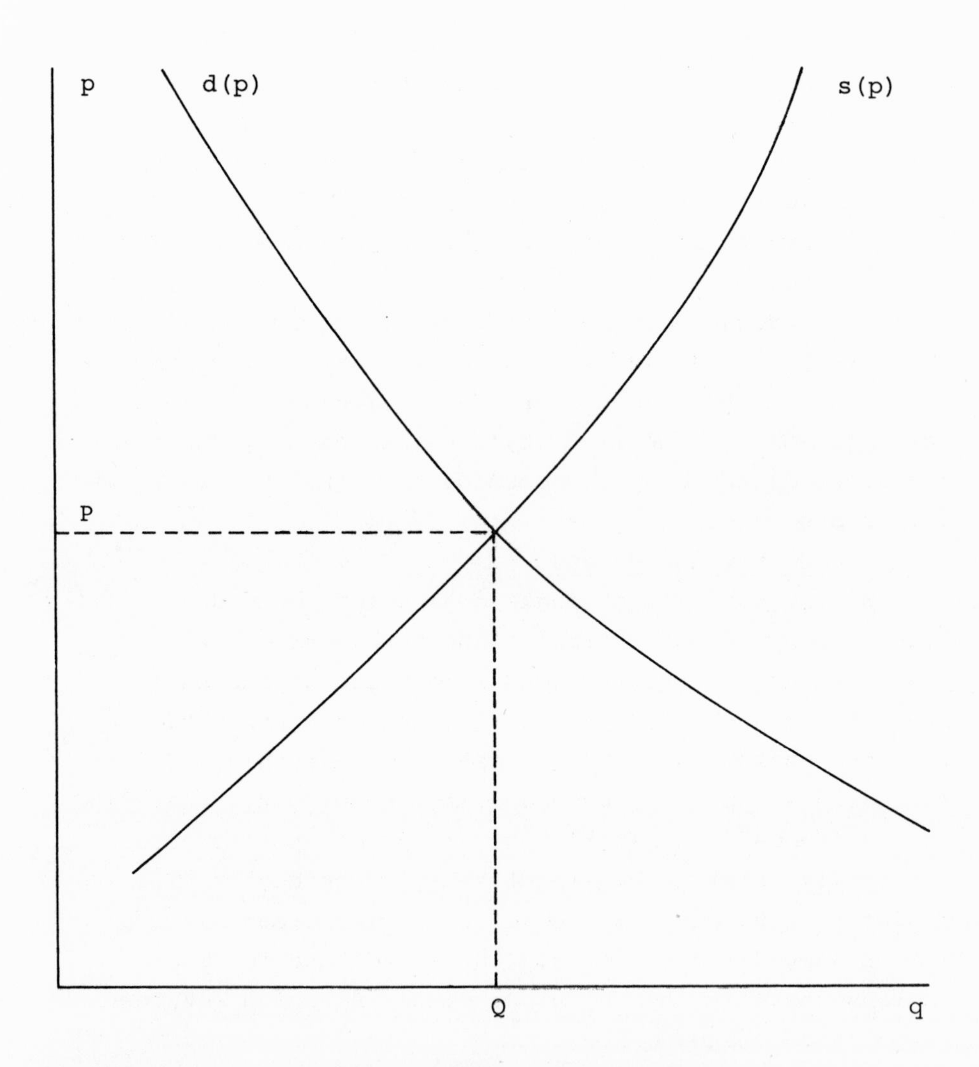

Supply and demand ("Marshall's scissors")

Now, for a moment, drop the assumptions that all transactions are at the same price and suppose, instead, that the consumers pay for each additional amount of the good the maximum price that they are willing to pay. Then the total they would pay for the quantity Q would be

$$\int_0^Q \pi(q)dq \,, \tag{2.2}$$

where $p = \pi(q)$ is the inverse function of $q = d(p)$. In the graphical representation (2.2) is the area under the curve $d(p)$. The integral (2.2) is in some sense the value of the quantity Q to the consumers. Now reverting to the original supposition that all transactions take place at price P, we see that the consumers have benefited by the consumers' surplus

$$\begin{aligned} &\int_0^Q \pi(q)dq - PQ \\ &= \int_0^Q \{\pi(q)-P\}dq \\ &\geq 0 \quad . \end{aligned} \tag{2.3}$$

Analogously there is the producers' surplus

$$\begin{aligned} &PQ - \int_0^Q \sigma(q)dq \\ &= \int_0^Q (P-\sigma(q))dq \\ &\geq 0 \,, \end{aligned} \tag{2.4}$$

where $p = \sigma(q)$ is the inverse of $q = s(p)$.

The preceding analysis (due to our Alfred Marshall) depends on the tacit assumption that (in defining $s(p)$, $d(p)$) we can change the price of our one good without affecting the rest of the economic situation - the changes are ceteris paribus (other things being equal). But in an interdependent economy one cannot change the price of one good while keeping everything else (prices, quantities traded, of other goods) all fixed. In his definition of $s(p)$, $d(p)$ Marshall did not specify which were the other things to be kept equal, and there is contention amongst economists as to what he meant, and what he should have meant. If the good under consideration plays a minor part in the economy (e.g. bubble-gum) then it is reasonable to assume that change in its price has only minimal repercussions and the above is meaningful; but if it plays a major part (e.g. raw steel) then in any case a more sophisticated analysis is required. (Even if we can attach a meaning to $\pi(q)$ for small q, the integral defining the consumers' surplus might well diverge.)

As a digression we now consider the effect of charging a tax $\delta > 0$ on the producer for every unit of the good produced. If p is the price paid by the consumer, the price received by the producer is now $p - \delta$. Hence the price P^* and quantity Q^* which equates supply and demand are now given by

$$d(P^*) = s(P^*-\delta) = Q^* . \tag{2.5}$$

Clearly

$$Q^* \le Q \tag{2.6}$$

and

$$P^*-\delta \le P \le P^* , \tag{2.7}$$

where the signs of equality can hold only on the extreme assumption that $s(p)$ or $d(p)$ is locally constant. Further, the cost of the tax is in effect divided between the producer (who now receives only $P^* - \delta$) and the consumer (who pays P^*) in a way which depends on the curves $s(p)$, $d(p)$ and which would have been precisely the same if the tax had been imposed on the consumer. If δ is small and $s(p)$, $q(p)$ have continuous derivatives, it is easy to verify that

$$P^* = P+s'\delta/(s'+|d'|) \tag{2.8}$$

$$P^*-\delta = P-|d'|\delta/(s'+|d'|) , \tag{2.9}$$

where s' is the derivative.

Finally, it is clear from a diagram that in the taxed situation the sum of the three items (i) consumers' surplus (ii) producers' surplus and (iii) the tax revenue δQ^* is less than the sum of (i) and (ii) in the untaxed situation.

3. Perfect competition

We consider an industry which manufactures a certain good. There are a large number of small firms which act entirely independently of each other. No single firm can make more than a very small proportion of the total amount of the good which is traded. Under these circumstances, the decision of the individual firm has only a negligible effect on the going price p_o of the good: if it tries to sell at a price $p > p_o$ it will find no buyers, but if it sells at a $p < p_o$ it can dispose of as much as it can make. In other words, on the scale of the individual firm the demand curve $d(p)$ is a horizontal line: $d(p) = 0$ for $p > p_o$ but $d(p) = \infty$ for $p < p_o$. (This is sometimes taken as the definition of a competitive industry.

For an individual firm let $C(q)$ be the cost of producing a quantity q of the good per unit time (year). We suppose that $C(q)$ is increasing and, where necessary, that it is continuously differentiable. There is a maximum amount q_∞ which the firm can produce so we put $C(q) = \infty$ for $q > q_\infty$. In general, the fixed cost $C(0)$ is strictly positive. We call $C(q)-C(0)$ the variable cost. We do not need to suppose that the function $C(q)$ is the same for all firms in the industry.

If the firm makes a quantity q and sells it at a price p it will make a profit

$$R(q,p) = pq - C(q) \tag{3.1}$$

if this is positive, or a loss $-R(q,p)$ if $R(q,p)$ is negative. The firm is supposed to choose q so that at the going price p_o it maximizes $R(q,p_o)$: that is, it makes as large a profit as it can or, if it cannot be profitable, it minimizes its loss.

We now look at the industry in the medium term. A firm which cannot break even will ultimately go out of business. On the other hand, if there is a firm which can make a strictly positive profit, it will be worth the while of an entrepreneur to build a carbon copy and enter the industry. The demise or entry of a single firm will have only a minimal effect on the going price p_o, by our hypothesis that all firms are small. If many firms go bust or are created, the going price p_o may change (being determined by demand and supply for the industry as a whole) but it is plausible that the industry will move towards a situation in which every firm breaks even but none can make a profit.

We have thus justified

Axiom 3.1. In a competitive industry with going price p_o an individual firm with cost function $C(q)$ manufactures a quantity $q_o > 0$ such that

$$p_o = C(q_o)/q_o = \inf_q C(q)/q . \qquad (3.2)$$

Further,

$$p_o = C'(q_o) . \qquad (3.3)$$

For (3.2) just expresses that $R(q,p_o) \le 0$, with equality at q_o; and (3.3) holds since $R(q,p_o)$ has a maximum at q_o.

Corollary. In a competitive industry the producers' surplus is 0.

The "invisible hand" of competition thus confers all the benefits upon the consumer. It is one of the guiding principles of economics that, even if an industry is not competitive in the sense defined above, it is working to the greatest benefit of the community if it satisfies (3.3):

Principle 3.1. Consider an industry producing a good

in response to demand. Let $C(q)$ be the cost of producing a quantity q and let $\pi(q)$ be the price at which demand consumes q (as in §2). Then the benefit of the community is maximized if the quantity q_o produced satisfies

$$\pi(q_o) = C'(q_o) \quad . \tag{3.4}$$

We have called this a "Principle" because the expression "benefit of the community" remains undefined and because if we do define it (in a variety of ways) we shall find that the conclusion of the Principle follows from our other assumptions.

We can justify the Principle in the context of §2 as follows. Suppose that the amount actually produced is Q . Then $\pi(Q)$ is the value of a further unit of the good. It costs $C'(Q)$ to produce. If $\pi(Q) > C'(Q)$ then it is worthwhile to produce the extra unit: but if $\pi(Q) < C'(Q)$ it would have been better to have produced a unit less. Alternatively, the net benefit to the community of producing Q units is

$$\int_o^Q \pi(q)dq - C(Q) \quad . \tag{3.5}$$

If this is maximized at $Q = q_o$, then (3.4) holds.

We digress to give a further situation in which Principle 3.1 is verified. Suppose that there are n goods $1,\ldots,n$ where good j is manufactured by an industry with cost function $C_j(q)$ $(1 \le j \le n)$. Suppose, further, that there is a utility function $u(\underline{q})$ where $\underline{q} = (q_1,\ldots,q_n)$ is any bundle of the commodities. Suppose that a total money resource M is available. Then there is greatest utility if

$$u(q_1,\ldots,q_n) \tag{3.6}$$

is maximized subject to

$$\Sigma\, C_j(q_j) = M \ . \tag{3.7}$$

Hence the optimum vector $\underline{q}^*$ satisfies

$$u_j(\underline{q}^*) = \lambda C_j'(q_j^*) \,, \tag{3.8}$$

where $u_j = \partial u/\partial q_j$ and λ is a Lagrange multiplier. Now suppose in a market economy that the goods are traded at a price vector $\underline{p}$. Then the consumer will in fact maximize subject to

$$\Sigma\, p_j q_j = M \,, \tag{3.9}$$

and so obtain a bundle $\underline{q}^o$ with

$$u_j(\underline{q}^o) = \mu p_j \tag{3.10}$$

for some multiplier μ. Now suppose that one of the industries, say the first, is competitive, and hence $p_1 = C_1'(q_1)$. Then we can only have $\underline{q}^o = \underline{q}^*$ if all the other industries have $p_j = C_j'(q_j)$; i.e. if they produce in accordance with Principle 3.1.

4. Monopoly

We now consider a good where there is a single supplier who has a (natural or legal) monopoly. We suppose that he is subject to the law of diminishing returns:

$$C'(q) > 0 \,, \; C''(q) > 0 \,, \tag{4.1}$$

or, equivalently, that $C(q)$ is increasing and convex. As before, $\pi(q)$ is the price at which the demand is q, and $\pi(q)$ decreases with q.

The profit on a production of q is thus

$$R(q) = q\pi(q) - C(q) \,: \tag{4.2}$$

and the monopolist is supposed to choose q so as to maximize this, say at q_m. (The hypotheses do not imply that q_m is unique.) Put

$$p_m = \pi(q_m) \,, \tag{4.3}$$

the monopolist's price. Equating the derivative of (4.2) to 0 we obtain

$$\begin{aligned} p_m &= C'(q_m) - q_m\pi'(q_m) \\ &> C'(q_m) \,, \end{aligned} \tag{4.4}$$

if we ignore the very exceptional possibility $\pi'(q_m) = 0$, i.e. that the demand curve is locally horizontal. The monopolist sells at above his marginal cost of production. If he were to lower his price he could sell more and still make a profit on the extra production: but this would be more than offset by the decreased profit on his original output q_m .

The competitive quantity q_c and price p_c are obtained by equating price to marginal cost:

$$p_c = \pi(q_c) = C'(q_c) \ . \tag{4.5}$$

Since $C'(q) - \pi(q)$ is an increasing function of q , on comparing (4.4) and (4.5) we obtain

$$q_m < q_c \ , \tag{4.6}$$

and hence

$$p_m > p_c \ . \tag{4.7}$$

Hence under monopoly the consumers' surplus is reduced, as also is the total benefit to society in the sense discussed at the end of the last section. The payment

$$q_m(p_m - p_c) \tag{4.8}$$

which the monopolist receives because he charges p_m rather than p_c is known (somewhat misleadingly) as his economic rent. A justification for the term is that economic rents tend in practice to be converted into actual rents which are part of the cost of production. For example, a manufacturer with a patent enjoys an economic rent. If, however, he permits other manufacturers to use the patent subject to royalty, then for them the royalty payment is just one of the costs of production.

It should perhaps be remarked explicitly that in the monopoly situation there is no reason to expect that the profit $R(q_c)$ is zero. Both $R(q_c) > 0$ and $R(q_c) < 0$ are compatible with our assumptions (and, presumably, can occur in practice). If $R(q_c) > 0$, the dirigiste might regard it as reasonable to compel the monopolist to increase production from q_m to q_c : but if $R(q_c) < 0$ a subsidy would also be required.

5. Duopoly

Here there are precisely two producers, labelled 1,2. Each has a convex cost function C_j as in §4. The total supply is

$$q = q_1 + q_2 , \tag{5.1}$$

where firm j produces q_j . The price which equates this to the demand is $\pi(q)$, where π is as in §4. The two firms thus obtain profits

$$R_j(q_1,q_2) = q_j\pi(q_1+q_2) - C_j(q_j) . \tag{5.2}$$

Cournot, who was the first to discuss the problem, assumed that each producer would adjust his output to maximize his profit on the assumption that the other producer would keep his output fixed. If this leads to a stable situation, it must satisfy

$$\partial R_j(q_1,q_2)/\partial q_j = 0 \qquad (j = 1,2) . \tag{5.3}$$

Cournot's solution assumes that business men are not very intelligent. If producer 1 makes a small change δ in q_1 he suffers only a loss $O(\delta^2)$, but in general he imposes a change of order δ in R_2 . If now producer 2 reacts in the Cournot way by adjusting q_2 to maximize R_2 , then (if δ was wisely chosen) producer 1 will be better off than at the Cournot point.

Later workers assume that the producers will somehow gravitate to the core. A point (q_1^*,q_2^*) is in the core if for every other (q_1,q_2)

$$R_j(q_1,q_2) < R_j(q_1^*,q_2^*) \tag{5.4}$$

for at least one j . There is a representation very similar to the Edgeworth box. If q_1,q_2 are plotted as cartesian coordinates, then (q_1^*,q_2^*) is in the core if the two curves

$$R_j(q_1,q_2) = \text{const.}, \tag{5.5}$$

which pass through it, touch. The points of the core

Figure 5.

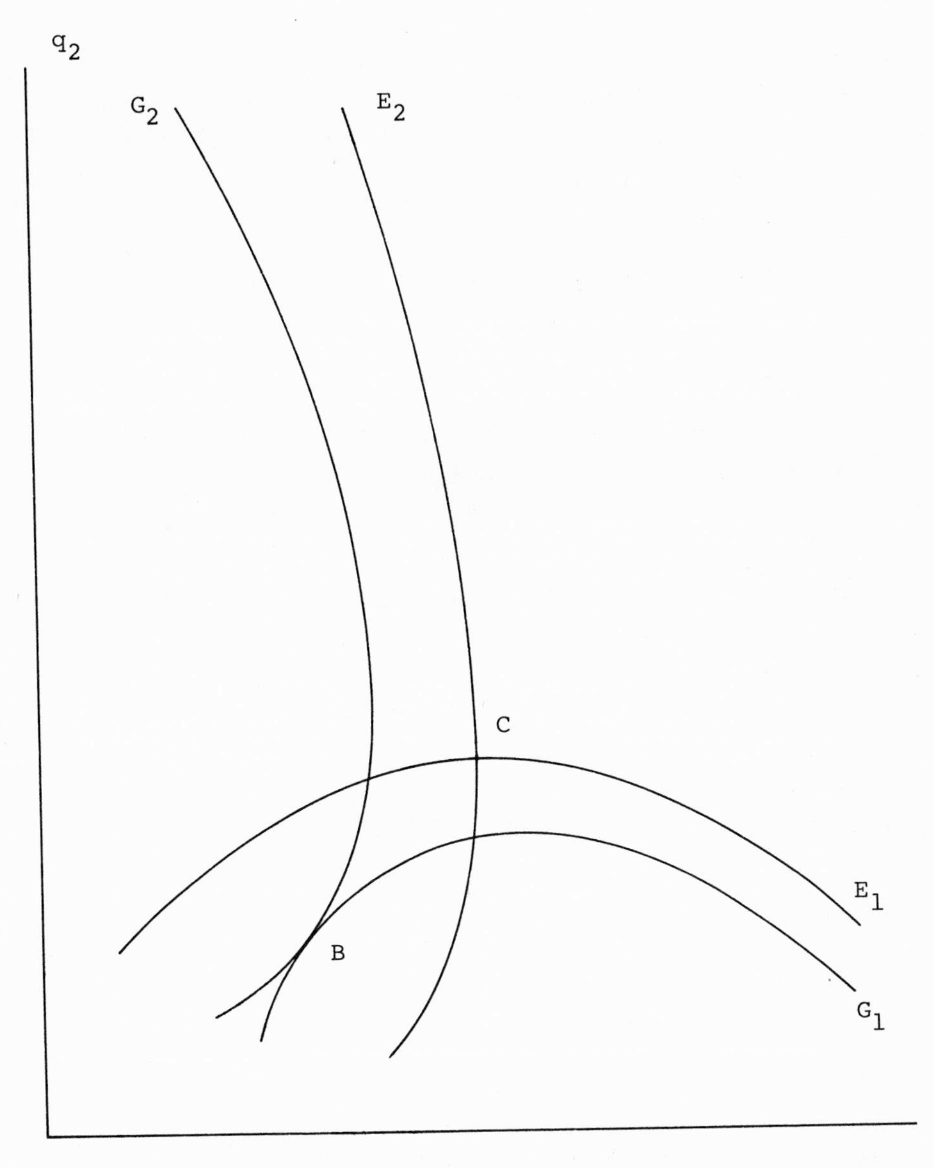

Duopoly. Here E_j, G_j are curves of equal profit for j (= 1,2). The point C (where one tangent is horizontal and the other vertical) is a Cournot point and B (where two curves touch) is a point of the core.

form a <u>contract curve</u>. The particular point on the contract curve which is attained is supposed to depend on the tactical skill of the producers and on adventitious circumstances in the process of reaching equilibrium.

If the two producers are able to cooperate, they have a monopoly position and will choose q_1, q_2 so as to maximize the total profit

$$R_1(q_1,q_2) + R_2(q_1,q_2) .$$

This gives a point of the core, but in general there will be others.

6. <u>Oligopoly</u>

This is when the number of producers is sufficiently small that the decision of an individual producer about outputs has an appreciable effect on the price which matches supply and demand. In general, the situation can be complicated, particularly if the producers are allowed to cooperate or to form coalitions to a greater or lesser extent. We mention two general points.

(i) Suppose that the producers are not allowed to cooperate. It is observed in practice that changes in price are infrequent in response to changes in demand or other circumstances. In other words, each producer acts as though his demand curve has a discontinuity of slope at the going price. If he were to raise his price, the other producers increase their market share by keeping their prices fixed. On the other hand, if he were to lower his price, the others would be forced to lower theirs so as to maintain their share, and the advantages would be diminished.

(ii) Suppose that there is <u>free entry</u> to the industry: that is, an entrepreneur is at liberty to build a factory and start manufacturing the good. This is certainly profitable if the going price is P and the cost function $C(q)$ of the new factory is such that it can make a profit on the portion of the demand curve which is not met at price P : that is, with a demand curve

$$d^*(p) = \begin{cases} 0 & (p \geq P) . \\ d(p) - d(P) & (p < P) . \end{cases} \quad (6.1)$$

It is in the interest of the existing producers to charge a price P sufficiently low that new entrants are inhibited. This limits the extent to which they can exploit their oligopolistic position. The phenomenon is known as stay-out pricing.

7. Factor costs

Hitherto, we have discussed the cost of producing a good as given in money terms. We now analyse the cost in terms of the various goods which may be used as inputs which we label $1,\ldots,n$. In general, a given quantity x of output may be obtained from several different bundles of inputs $\underline{y}$ (more labour or greater use of materials, the mix of skilled and unskilled labour, different proportions of alloys in steel, etc.). The possibility set Y is the set of

$$(x,\underline{y}) \in R^{n+1} \quad (7.1)$$

such that the given firm can produce x of output from the bundle $\underline{y}$ of inputs. It is usual to suppose, at least for most industries, that the possibility set Y is convex (Law of diminishing returns in the substitution of one input for another) and we shall assume this. [If Y is not assumed convex then some of the argument can be modified by using its convex hull.] A point $(x,y) \in Y$ is efficient if there is no $\underline{y}^* < \underline{y}$ for which $(x,\underline{y}^*) \in Y$. We suppose that the firm chooses only efficient points and that these lie on a hypersurface

$$x = g(\underline{y}) . \quad (7.2)$$

Clearly x is an increasing function of $\underline{y}$. The hypothesis that Y is convex implies that $g(\underline{y})$ is a concave function (i.e. that $-g(\underline{y})$ is convex). In this section we suppose that $g(\underline{y})$ is strictly concave. This implies, in particular, that for any $\underline{y} > \underline{0}$ the function $g(\lambda\underline{y})$ decreases with λ, that is that there are decreasing returns to scale. The case of constant returns to scale (i.e. $g(\underline{y})$ homogeneous of degree 1) requires some modifications of the argument but can be dealt with similarly. If there are economies of scale (increasing returns to scale) then the possibility set Y is no longer convex and the situation becomes entirely different (cf. Chapter 4, §4). We shall also assume that $g(\underline{y})$ is continuously differentiable to the extent that the context requires.

Suppose that the price r of the output and the prices $\underline{p}$ of the inputs are given. If the firm produces x output from inputs $\underline{y}$ then its profit (ignoring fixed costs) is

$$rx - \underline{p}\underline{y} \; . \tag{7.3}$$

The firm maximizes this, say at $(\xi, \underline{\eta})$. Put

$$R = R(r,\underline{p}) = r\xi - \underline{p}\underline{\eta}. \tag{7.4}$$

Clearly the hyperplane

$$rx - \underline{p}\underline{y} = R \tag{7.5}$$

is a tac-hyperplane to $x = g(\underline{y})$ at $(\xi, \underline{\eta})$; we have

$$\xi = g(\underline{\eta}) \tag{7.6}$$

(the point $(\xi, \underline{\eta})$ is efficient); and

$$g_j(\underline{\eta}) = p_j/r \; , \tag{7.7}$$

at least if $\eta_j > 0$. Here, as usual,

$$g_j(\underline{y}) = \partial g/\partial y_j \; . \tag{7.8}$$

The equation (7.7) has an obvious economic interpretation. The function $g_j(\underline{y})$ is the marginal productivity of the input j .

The investigation of the effect of changes in the prices r, $\underline{p}$ is very similar to the investigation of an indifference hypersurface in Chapter 1, §3, so we can be brief. Let $(r^*,\underline{p}^*)$ and $(r^o,\underline{p}^o)$ be two sets of prices and let $(\xi^*,\underline{\eta}^*)$, $(\xi^o,\underline{\eta}^o)$ be the corresponding values of $(\xi,\underline{\eta})$. The definition of $(\xi,\underline{\eta})$ implies that

$$r^*\xi^o - \underline{p}^*\underline{\eta}^o \leq r^*\xi^* - \underline{p}^*\underline{\eta}^* , \qquad (7.9)$$

$$r^o\xi^* - \underline{p}^o\underline{\eta}^* \leq r^o\xi^o - \underline{p}^o\underline{\eta}^o ; \qquad (7.10)$$

and hence

$$(r^*-r^o)(\xi^*-\xi^o) - (\underline{p}^*-\underline{p}^o)(\underline{\eta}^*-\underline{\eta}^o) \geq 0 . \qquad (7.11)$$

In particular, on varying only one of r or p_j at a time, (7.11) implies that

$$\partial\xi/\partial r \geq 0 ; \ \partial\eta_j/\partial p_j \leq 0 , \qquad (7.12)$$

whenever the differential coefficients exist. There is an obvious economic interpretation.

If R is defined by (7.4), we have

$$dR = (rd\xi - \underline{p}d\underline{\eta}) + (\xi dr - \underline{\eta}d\underline{p})$$

$$= \xi dr - \underline{\eta}d\underline{p} , \qquad (7.13)$$

since (7.5) is a tac-hyperplane. This is a perfect differential, so

$$\partial\eta_i/\partial p_j = \partial\eta_j/\partial p_i , \qquad (7.14)$$

and

$$\partial\eta_j/\partial r = -\partial\xi/\partial p_i . \qquad (7.15)$$

If the common value (7.14) is positive, then inputs i,j are <u>substitutes</u>, otherwise they are <u>complements</u>.

By (7.4) we have

$$\partial\xi/\partial r = \sum_j g_j(\underline{\eta})\partial\eta_j/\partial r ; \qquad (7.16)$$

and so

$$\partial\eta_j/\partial r \geq 0 \qquad (7.17)$$

for at least one j , by (7.12). Again there is an obvious interpretation. One can however construct examples (even with $n = 2$) when (7.17) does not hold for all j : which is, perhaps, counter-intuitive.

In conclusion, we note that the analysis can readily be extended to cover the case of an industry which produces several outputs, whose prices can vary independently (e.g. the sheep industry produces wool and mutton) on the assumption that the production set is convex. It is convenient to take the hypersurface of efficient points in the shape

$$f(\underline{x},\underline{y}) = 0 ,$$

where $\underline{x}$ is the bundle of outputs and $\underline{y}$ that of inputs. The details are left to the reader.

Chapter 3 Exercises

1. The cost of running the rail service from Cambridge to London is independent of the number of passengers. The railway company fixes the price of first and second-class tickets so as to maximize its revenue. Consumer research indicates that there are four types of potential passenger. Their numbers and conditions for travelling first and second class are given below, where p_j is the price of a j-class ticket (in £).

Class	Number	Will travel First if	Will travel Second if
A	450	$p_1 \leq p_2+4$ and $p_1 \leq 12$	Not First and $p_2 \leq 7$
B	40	$p_1 \leq 11$	Never
C	900	$p_1 \leq p_2 \leq 8$	Not First and $p_2 \leq 8$
D	200	$p_1 \leq p_2 \leq 6$	Not First and $p_2 \leq 6$

(i) Show that the Railway will choose $p_1 = 12$, $p_2 = 8$.

(ii) A Labour government imposes a tax of £7 on a First-class ticket but no tax on a Second-class ticket. Show that the Railway will reduce the price of both First and Second-class tickets, to $p_1 = 11$, $p_2 = 6$ respectively.

[Edgeworth Paradox. Hotelling, *J.Polit.Econ.* 40

(1932), 577-616.]

2. In the notation of §4 examine whether a monopolist can be induced to adopt the competitive level of output by imposing a tax $A + Bq$ where q is his output and A,B are appropriately chosen constants.

3. Two duopolists are engaged in the production of mineral water. It costs nothing to produce. The price at which it can be sold is $\pi(q) = 1-q$, where $q = q_1 + q_2$ and q_1, q_2 are the quantities produced by the two duopolists.

 (i) Find their total profit if they combine and act monopolistically.

 (ii) Find the profit of each at the Cournot point.

 (iii) Find the "contract curve".

4. ("Le Chatelier principle".) Notation as in §7. Let $(\xi^*, \underline{\eta}^*)$ correspond to prices $(r^*, \underline{p}^*)$. Consider also the situation M in which the price p_n and the quantity y_n of the n-th good are constrained by $p_n = p_n^*$, $y_n = \eta_n^*$. Show that

$$(\partial\xi/\partial r) \geq (\partial\xi^M/\partial r) \geq 0 ,$$

where M refers to the modified situation, and where both derivatives are taken at $(r,\underline{p}) = (r^*,\underline{p}^*)$. [Hint. show that $R^M(r,\underline{p}) \leq R(r,\underline{p})$ on $p_n = p_n^*$, and consider first and second derivatives with respect to r. References: Samuelson Collected Papers vol.1 and his Nobel Prize speech, Amer.Econ. Rev. 62 (1972), 249-262.]

5. In the notation of §7 construct an example with $n = 2$ for which

$$\partial\xi/\partial p_1 > 0 > \partial\xi/\partial p_2 .$$

6. (i) An industry A produces n goods ("intermediates") which are used by a second industry B to produce a consumption good whose price

is fixed. The intermediates are neither used nor produced by any other industry. Suppose that the cost to A of producing a bundle $\underline{x}$ of intermediates is $v(\underline{x})$, where v is convex (e.g. $v(\underline{x}) = \Sigma C_j(x_j)$, where the C_j are convex). Suppose industry B produces consumption good of value $u(\underline{x})$ from bundle $\underline{x}$, where u is concave. Let $\underline{p}$ be the prices at which industry A sells the intermediates to industry B, so that they maximize

$$R_A = \underline{px} - v(\underline{x}) \ , \ R_B = u(\underline{x}) - \underline{px}$$

respectively. Show that supply and demand lead to the selection of a bundle $\underline{y}$ of intermediates which maximizes

$$u(\underline{y}) - v(\underline{y}) \ ;$$

and that

$$p_j = u_j(\underline{y}) = v_j(\underline{y}) \ ,$$

where u_j, v_j are partial derivatives.

(ii) Suppose that government exacts a tax t_j on intermediate j $(1 \leq j \leq n)$. Show that the industries will choose a bundle of intermediates $\underline{z}$ to maximize

$$u(\underline{z}) - v(\underline{z}) - \underline{tz} \ .$$

Show, further, that the sum

profits of A + profits of B + tax revenue $\underline{tz}$

is <u>always</u> strictly less than the sum of the profits of A and B in the untaxed situation. Show that the result continues to hold if some or all of the intermediates are subsidized rather than taxed.

(iii) Suppose that government imposes a Value Added Tax at the same rate on both industries. Show that the total of tax revenue and the profits of A and B is equal to the sums of the profits of A and B in the untaxed situation.

(iv) Suppose that there are 2 intermediates.

Show that the imposition of a small (infinitesimal) tax on the first intermediate (but no tax on the second) may either raise or lower the price of the second intermediate. Find conditions on the first and second derivatives of u, v at $\underline{y}$ which ensure that the second price is lowered. Consider also the first price (both before and after tax). [Hotelling. J. Polit. Econ. 40 (1932), 577-616. For (ii) cf. §2: for (iv) cf. Exercise 1.]

7. The *elasticity* e at x_o of a function $f(x)$ is defined to be $|d \log f(x)/d \log x|$ at x_o. Further, $f(x)$ is *elastic* or *inelastic* according as $e > 1$ or $e < 1$. If the demand for a good at price p is $d(p)$, show that the revenue $pd(p)$ increases or decreases as p increases according as whether $d(p)$ is inelastic or elastic.

8. Consider the following simplified version of international trade. There are two countries A, B whose units of currency are called respectively the alpha and the beta. The rate of exchange is ρ alphas for one beta. Good 1 is produced only in A and consumed only in B; contrariwise good 2 is produced only in B and consumed only in A.

   (i) Suppose that A can produce any quantity of good 1 at a price of p_1 alphas a tonne. Let ε_1 be the price-elasticity of demand for good 1 in country B (where the price, of course, is p_1/ρ betas). Similarly B can produce any quantity of good 2 for p_2 betas a tonne and ε_2 is the price-elasticity of demand for good 2 in A. Suppose, also, that the trade is in equilibrium: that is

$$p_1X_1 = \rho p_2X_2 ,$$

where X_j is the amount of good j traded. If now the rate of exchange becomes $\rho^* = \rho(1+\delta)$, where $\delta > 0$ is small, show that country A has a surplus or deficit in the trade according as $\varepsilon_1 + \varepsilon_2 > 1$ or $\varepsilon_1 + \varepsilon_2 < 1$.

[Note. This is known as the Marshall-Lerner condition.]

(ii) Instead of supposing that A produces any quantity of good 1 at a fixed price, suppose that there is a price-elasticity η_1 of production. Show that the price (in alphas) at which good 1 is traded has an elasticity $\varepsilon_1/(\varepsilon_1+\eta_1)$ as a function of the exchange rate ρ, and find the elasticity of the quantity X_1 traded.

(iii) As in (i), except that good j has elasticity η_j of production. Show that, when ρ increases slightly, the country A moves into surplus or deficit according as $\lambda_1+\lambda_2 > 1$ or $\lambda_1+\lambda_2 < 1$, where $\lambda_j = \varepsilon_j(1+\eta_j)/(\varepsilon_j+\eta_j)$.

[Note. (i) is the case $\eta_1 = \eta_2 = \infty$.]

9. (tedious but not difficult). A firm makes a good Z from outputs X,Y. Suppose that the quantity z of Z obtained by use of quantities x,y of X,Y is

$$z = f(x,y)$$

where there are constant returns to scale (i.e. $f(x,y)$ is homogeneous of degree 1) but $f(x,y)$ is strictly concave on $x+y = 1$. Let X,Y,Z have prices p,q,r respectively.

(a) For given p,q show that the firm will adopt a fixed ratio x/y of inputs. Show, further, that the elasticity of x/y in terms of p/q is given by

$$\sigma = \frac{f_1(x,y)f_2(x,y)}{f(x,y)f_{12}(x,y)}, \qquad \text{(i)}$$

where the subscripts denote partial derivatives. Show also that

$$p = rf_1 \; ; \; q = rf_2 \; . \qquad \text{(ii)}$$

(b) Suppose that the industry is competitive. Show that

$$rz = px + qy \; . \qquad \text{(iii)}$$

Under small changes of demand or supply (affecting p, q, r, x, y, z but not the function f) show that

$$rdz = pdx + qdy \qquad \text{(iv)}$$

and

$$zdr = xdp + ydq \; . \qquad \text{(v)}$$

Now suppose that the price-elasticity e of supply of Y is given. We want to find the price-elasticity of demand λ of X in terms of the price-elasticity of demand η of Z.

(c) Deduce from (v) that

$$\eta^{-1} rdz = \lambda^{-1} pdx - e^{-1} qdy \; . \qquad \text{(vi)}$$

(d) By differentiating the second equation in (ii), show that

$$\eta^{-1} rdz = \sigma^{-1} pdx - \theta qdy \qquad \text{(vii)}$$

where

$$(1-\kappa)\theta = 1/e + \kappa/\sigma \; , \qquad \text{(viii)}$$

$$\kappa = px/rz \qquad \text{(ix)}$$

and σ is given by (i).

(e) Deduce from (iv), (vi), (vii) that

$$\lambda = \{\sigma(\eta + e) + \kappa e(\eta - \sigma)\}/\{\eta + e - \kappa(\eta - \sigma)\} \; . \qquad \text{(x)}$$

(f) According to Marshall and Pigou:

"I. The demand for anything is likely to be more elastic, the more readily substitutes for that thing can be obtained.

II. The demand for anything is likely to be less elastic, the less important is the part played by the cost of that thing in the total cost of some other thing, in the production of which it is employed.

III. The demand for anything is likely to be

more elastic, the more elastic is the supply of cooperant agents of production."

IV. The demand for anything is likely to be more elastic, the more elastic is the demand for any further thing which it contributes to produce."

Interpreting this to mean that λ increases with σ, κ, e, η respectively, show that I,III,IV are true in this model but that II is true precisely when $\eta > \sigma$.

[Hicks, The theory of wages.]

CHAPTER 4. WELFARE ECONOMICS

1. Introduction

Hitherto we have considered the actions and interactions of agents (persons, firms) each of which seeks to maximize its own utility. This may lead to situations which (by fairly general consent) are not optimal for society as a whole. Any attempt to formalize such goals as "the greatest happiness of the greatest number" leads to comparisons of the utilities of different persons. More generally, some economists have introduced a social utility function which measures the welfare of society as a whole (or of a specified subsociety, e.g. a family). This may depend only on the utilities of individuals (e.g. their sum) or may be more complicated in structure.

The notion of social utility is distinctly more rarified than that of an individual's utility. It is perhaps not too grossly unrealistic to suppose that an individual, confronted with a number of alternatives, can put them in order of preference. It is much more questionable to think of an order decided by a society as a whole. Indeed, "Arrow's impossibility theorem" shows, subject to some very innocuous-seeming axioms, that it is impossible to construct a social ordering from the preference orderings of the individuals in the society.

Nevertheless, one may suppose that, by some sort of consensus or otherwise, a social utility function is given. An individual agent acts, however, to increase its own utility and this may well decrease social utility. It may be regarded as a duty of Government to maximize social utility either by direct action or by manipulating market forces. Here it has to be borne in mind that government action may have its costs (e.g. employment as tax collectors or safety inspectors of people who could otherwise be productively employed); and these have to be taken into account in assessing the effect of government action on social utility.

Rather than pursue these generalities further, we discuss a number of simple concrete situations. At the end we prove a form of Arrow's impossibility theorem.

2. Public good

This is a good which benefits all, whether or not they have paid towards its production (e.g. weather forecasts). The problem of ensuring that all contribute fairly to the cost of producing an optimum amount of public good is picturesquely called the free rider problem.

We take a very crude model. Consider two goods 1 a public good and 2 a private good. Each household has a utility function

$$u_h(X, y_h) , \tag{2.1}$$

where X is the amount of the public good and y_h is the household's quantity of the private good. Let p_1, p_2 be the prices of the two goods (supposed independent of the demand), and suppose that each household h has a budget B_h from which it makes a contribution x_h. Then

$$X = \Sigma x_h \tag{2.2}$$

and

$$p_1 x_h + p_2 y_h = B_h . \tag{2.3}$$

The individual household h may be expected to pick x_h, y_h subject to the budget constraint (2.3) so as to maximize (2.1) under the supposition that other households $k \neq h$ will keep their contributions unaltered.

It will therefore select x_h, y_h so that

$$u_{h1}(X,y_h) : u_{h2}(X,y_h) = p_1 : p_2 , \tag{2.4}$$

where u_{h1}, u_{h2} are the partial derivatives.

This will, however, mean that too little of good 1 is produced. To avoid the introduction of a social utility function, let us suppose for simplicity that all households have the same utility function and the same budget constraint, and also that they all contribute the same amount x of the public good. Then each will have a utility

$$u(Nx,y) , \tag{2.5}$$

where N is the number of households and

$$p_1x + p_2y = B . \tag{2.6}$$

Now maximization of (2.5) subject to (2.6) gives

$$u_1(Nx,y) : u_2(Nx,y) = p_1/N : p_2 . \tag{2.7}$$

The individual household should thus be induced to act as though the price of the public good is p_1/N .

3. Service subject to congestion

We consider a good where the use by one household reduces the utility of the good to other households (e.g. telephone conversations, where congestion makes the service less efficient).

Let 1 be a good of this type and let 2 be a normal good (or, say, money). We suppose that the utility to household h of quantities x_h, y_h of goods 1,2 respectively is

$$u_h(x_h, y_h, X) , \tag{3.1}$$

where

$$X = \sum_k x_k \tag{3.2}$$

is the total amount of good 1. Here (3.1) is a decreasing function of X (for fixed x_h, y_h), but for fixed X it is a utility function of the usual type.

Let p_1, p_2 be the costs of production of goods 1,2,

assumed independent of the quantities consumed. The individual h will regard X as fixed, and so will allocate a budget B_h between goods 1,2 so that

$$p_1x_h + p_2y_h = B_h \tag{3.3}$$

and

$$u_{h1}(x_h,y_h,X) : u_{h2}(x_h,y_h,X) = p_1 : p_2 \tag{3.4}$$

with our usual convention on partial derivatives.

From a social standpoint (3.4) implies that every household will demand too much of good 1. To see this, let us assume that all households h have the same utility function $u(x,y,X)$ and the same budget B, and that they purchase the same quantities x,y of goods 1,2. Then the utility of the individual household is

$$u(x,y,Nx) , \tag{3.5}$$

where N is the number of households; and this is to be maximized subject to

$$p_1x + p_2y = B . \tag{3.6}$$

The condition for maximum is now

$$u_1 + Nu_3 : u_2 = p_1 : p_2 \tag{3.7}$$

where u_1,u_2,u_3 are the partial derivatives of $u(x,y,X)$ with respect to x,y,X respectively. Let the maximum be attained at x^*,y^* and let u_j^* be the corresponding values of the derivatives. The individual household will select x^*,y^* if it makes its choice under prices p_1^*,p_2 such that

$$u_1^* : u_2^* = p_1^* : p_2 \tag{3.8}$$

and the budget constraint is appropriately modified. Comparison of (3.7) and (3.8) gives

$$\begin{aligned} p_1^* &= p_1u_1^*/(u_1^* + Nu_3^*) \\ &= p_1 + \delta \text{ (say)} > p_1 ; \end{aligned} \tag{3.9}$$

and x^*,y^* lies on the budget line

$$p_1^*x + p_2y = B+b , \tag{3.10}$$

with

$$b = \delta x^* > 0 . \tag{3.11}$$

Hence the individual will make the socially optimal choice if he is charged an inflated price $p_1^* > p_1$ for good 1 and given a rebate $b > 0$ independent of his choice.

In a real world the sums b would not necessarily be returned to the households: they might be used to improve the telephone service.

In the foregoing we have assumed, for simplicity, that the prices of goods 1,2 are independent of the amounts consumed. It is not difficult to modify the argument so as to take them to be the marginal costs of production for convex cost functions $C_1(X)$, $C_2(Y)$ with $Y = \sum_h y_h$.

4. Increasing returns to scale

Monopolies have already been discussed in an earlier chapter. They naturally arise in industries where there are increasing returns to scale, that is where the cost $C(x)$ of an individual firm of producing quantity x is such that $C(x)/x$ decreases as x increases. This may be because $C'(x)$ is decreasing or (as a limiting case) because there is a fixed cost $A > 0$ and there are constant variable costs to scale, so $C(x) = A+Bx$ for some $B > 0$. Here any situation with more than one firm is unstable: a firm with a larger market share than another can charge lower prices and so further increase its share until it has a monopoly.

The discussion of monopolies in Chapter 3 did not depend on the assumption made in that chapter that $C(x)$ is convex. The conclusion that to maximize social utility such a firm should equate its supply x to demand by charging a price $C'(x)$ holds in the more general situation.

5. Externalities

It sometimes happens that the production of one good has effects on the production of another good by an unrelated firm. This effect may be either malign (e.g. fumes from a chemical factory affecting local agriculture) or benign (e.g. effect of fruit production on local apiaries). Such an effect is called an externality. In the presence of externalities the maximization of profit by each firm individually does not lead to the greatest social benefit.

Consider two firms. The first produces good 1 and its cost of producing quantity x_1 is

$$C_1(x_1) \ , \tag{5.1}$$

where C_1 is a convex function of the usual type. The second firm produces good 2 and is affected by an externality from the first firm. Its cost of producing quantity x_2 (when the first firm is producing x_1) is

$$C_2(x_1,x_2) \ . \tag{5.2}$$

For fixed x_1 this is a convex function of the usual type; and we shall assume that

$$C_{21} = \partial C_2/\partial x_1 \tag{5.3}$$

is of constant sign (positive for a malign and negative for a benign externality).

We suppose that the two firms operate in a market where the prices of the goods are p_1,p_2 respectively (independent of x_1,x_2) .

Suppose, first, that the firms act entirely independently. Then the first firm maximizes

$$R_1(x_1) = p_1x_1 - C_1(x_1) \ , \tag{5.4}$$

and so will produce a quantity x_1^* given by

$$C_{11}(x_1^*) = p_1 \ . \tag{5.5}$$

Similarly the second firm will maximize

$$R_2(x_1,x_2) = p_2x_2 - C_2(x_1,x_2) \ , \tag{5.6}$$

given that $x_1 = x_1^*$, and so will choose x_2^* by

$$C_{22}(x_1^*,x_2^*) = p_2 \ . \tag{5.7}$$

Before discussing whether the choice of x_1^*,x_2^* is optimal for society, we note that it is not optimal for the two firms. If they cooperate to maximize the total revenue

$$R_1(x_1) + R_2(x_1,x_2) \ , \tag{5.8}$$

then they would agree to produce quantities x_1^o,x_2^o such that

$$C_{11}(x_1^o) + C_{21}(x_1^o,x_2^o) = p_1 \ , \tag{5.9}$$

$$C_{22}(x_1^o,x_2^o) = p_2 \ . \tag{5.10}$$

The second firm (the one affected by the externality) still equates marginal cost to price, but the first firm no longer does so. If $C_{21} > 0$ we have $C_{11}(x_1^o) < C_{11}(x_1^*)$ and so $x_1^o < x_1^*$: that is, the first firm is producing less because of the malign externality on the second firm. Similarly $C_{21} < 0$ gives $x_1^o > x_1^*$.

Even if we do not assume that the two firms will cooperate completely, we note that it will pay the second firm to bribe the first to produce a quantity different from x_1^* . If δ is small, then $R_1(x_1^*+\delta)$ differs from $R_1(x_1^*)$ only by a quantity of the second order, whereas $R_2(x_1^*+\delta,x_2^*)$ differs approximately from $R_2(x_1^*,x_2^*)$ by $-C_{21}(x_1^*,x_2^*)\delta > 0$ for δ of the appropriate sign. Hence the second firm can more than compensate the first firm for the change to $x_1^*+\delta$ and still have an increased revenue.

Somewhat paradoxically in view of what has been said elsewhere about the virtues of competition, we shall argue that social utility is maximized when the quantities produced are (x_1^o,x_2^o) . The argument is similar to that used to discuss monopolies. For comparison we introduce a third good whose production has no effect on and is unaffected by the production of the first two goods. It has a cost function

$$C_3(x_3) \tag{5.11}$$

of the conventional type. We abandon the assumption that the prices p_1, p_2 are fixed and seek to maximize a utility function

$$u(x_1, x_2, x_3) \tag{5.12}$$

for given budget

$$C_1(x_1) + C_2(x_1, x_2) + C_3(x_3) = B \; . \tag{5.13}$$

This will occur at $\bar{x}_1, \bar{x}_2, \bar{x}_3$ where

$$u_j(\bar{x}_1, \bar{x}_2, \bar{x}_3) = \theta \bar{p}_j \qquad (j = 1,2,3) \tag{5.14}$$

for some θ and with

$$\bar{p}_1 = C_{11}(\bar{x}_1) + C_{21}(\bar{x}_1, \bar{x}_2) \tag{5.15$_1$}$$

$$\bar{p}_2 = C_{22}(\bar{x}_1, \bar{x}_2) \tag{5.15$_2$}$$

$$\bar{p}_3 = C_{33}(\bar{x}_3) \; . \tag{5.15$_3$}$$

Here $\bar{p}_3$, the marginal cost of the third good, is equal to its price (in a competitive market). Hence $\bar{p}_1, \bar{p}_2$ are the prices of the first two goods; and, on putting $\bar{p}_j = p_j$, we see that $\bar{x}_j = x_j^o$ by comparison of (5.9), (5.10) and 5.15).

We have conducted the argument so far in terms of single firms producing goods 1 and 2, and so supposed that they could negotiate with one another. When there are large numbers of firms of each type this may no longer be possible: the "transaction costs" in securing the agreement of all the firms might well outweigh the benefits. Government could then attempt to secure the socially optimum production of the first good either by regulation or by imposing a tax (if the externality is malign) or a bounty (if benign). But, again, the additional costs of administration might not be worthwhile.

6. Arrow's impossibility theorem

Let S be a finite set. As usual, we define an ordering $\succ$ on S by transitivity together with the condition that precisely one of

$$x \succ y \ , \ y \succ x \ , \ x \asymp y \tag{6.1}$$

holds for every pair $x, y \in S$. Here we shall be concerned with several orderings on the same set and write

$$x \succ y \ (\phi) \tag{6.2}$$

to mean that $x \succ y$ in the ordering ϕ.

We consider a finite set (society) U of individuals i, each of which has a (preference) ordering ϕ_i on S. Arrow's theorem is concerned with finding a "social" ordering

$$\Phi = \Phi(\{\phi_i\}_{i \in U}) \tag{6.3}$$

which could be regarded as the ordering of S by U.

The following conditions appear natural.

I. Φ is defined when each of the ϕ_i independently runs through all orderings of the set S.

II. If $$x \succ y \ (\phi_i) \tag{6.4}$$
for all i, then
$$x \succ y \ (\Phi) \ . \tag{6.5}$$

III. (Indifference to irrelevant alternatives.) Let T be a subset of S. If for each i ϕ_i^{o} and ϕ_i^{*} induce the same ordering on T (which may depend on i), then so do $\Phi(\{\phi_i^{o}\})$ and $\Phi(\{\phi_i^{*}\})$.
We shall use III only when T has two elements, but it is easy to see that this implies III for general T.

[Note. It is not postulated that the ϕ_i occur symmetrically in (6.3).]

Theorem (Arrow). Suppose that S has at least 3 elements and that conditions I, II, III are satisfied. Then there is a $o \in U$ such that

$$\Phi(\{\phi_i\}) = \phi_o \ . \tag{6.6}$$

It is usual to say that the individual O is a <u>dictator</u>. The impossibility theorem says that there is no non-dictatorial social ordering Φ satisfying I,II,III. Arrow's original conditions were rather different: the above is a later version of his which is given in Amartya Sen's book.

Let Φ satisfy the hypotheses of the Theorem. We say that a subset $V \subset U$ is <u>almost decisive</u> for the ordered pair (x,y) with $x \neq y$ if

$$x \succ y \ (\phi_i) \ (i \in V) \ ; \ y \succ x \ (\phi_i) \quad (i \notin V) \tag{6.7}$$

implies

$$x \succ y \ (\Phi) \ . \tag{6.8}$$

<u>First step</u>. There is an individual $O \in U$ and a pair (x,y) such that (the set consisting of the single individual) O is almost decisive for (x,y).

For there certainly are sets V and pairs (x,y) such that V is almost decisive for (x,y), since U is almost decisive for any (x,y) by condition II. We choose V and (x,y) so that the number of elements of V is minimal. If V is a singleton set, we are done. Otherwise, V is the union of disjoint non-empty sets V_1, V_2. Let z be any third element of S, and choose orderings ϕ_i such that

$$x \succ y \succ z \ (\phi_i) \quad (i \in V_1) \tag{6.9$_1$}$$

$$z \succ x \succ y \ (\phi_i) \quad (i \in V_2) \tag{6.9$_2$}$$

$$y \succ z \succ x \ (\phi_i) \quad (i \notin V) \ . \tag{6.9$_3$}$$

Then

$$x \succ y \quad (\Phi) \tag{6.10}$$

since V is almost decisive for (x,y). By the minimality of V, the set V_2 is not almost decisive for (y,z), and so

$$y \not\succ z \quad (\Phi) \ . \tag{6.11}$$

(Note that we have here tacitly used condition III.) Similarly

$$z \not\succ x \quad (\Phi) \ . \tag{6.12}$$

But (6.10), (6.11) are together incompatible with Φ being an ordering. This completes the first step.

We shall say that O is <u>decisive</u> for the ordered pair (x,y) if

$$x \succ y \quad (\phi_o) \Rightarrow x \succ y \quad (\Phi) \tag{6.13}$$

and write $B(x,y)$. The statement that O is almost decisive for (x,y) is written $A(x,y)$. Clearly

$$B(x,y) \Rightarrow A(x,y) . \tag{6.14}$$

<u>Second step</u>. Let x,y,z be distinct. Then

$$A(x,y) \Rightarrow B(x,z) . \tag{6.15}$$

For consider any ϕ_i such that

$$x \succ y \succ z \quad (\phi_o) \tag{6.16$_1$}$$

$$y \succ x , \; y \succ z \quad (\phi_i) \quad (i \neq O) . \tag{6.16$_2$}$$

Then $A(x,y)$ implies $x \succ y$ (Φ) and condition II implies $y \succ z$ (Φ). Hence $x \succ z$ (Φ). The second step now follows by an appeal to condition III, on noting that (6.16$_2$) makes no statement about the relative positions of x,z under ϕ_i $(i \neq O)$.

<u>Third step</u>. If x,y,z are distinct, then

$$A(x,y) \Rightarrow B(z,y) .$$

The proof is similar.

We are now in business. The first step asserts that $A(x,y)$ is true for some (x,y). Let z be any third element. Then

$$A(x,y) \Rightarrow B(x,z) \Rightarrow B(x,y) ,$$

the second implication being by (6.14) and (6.15) with (x,y,z) replaced by (x,z,y). Further,

$$A(x,y) \Rightarrow B(x,z) \Rightarrow B(y,z) \Rightarrow B(y,x) .$$

Hence the true statement $A(x,y)$ implies $B(u,v)$ for any $(u,v) \in \{x,y,z\}$. It is now straightforward to deduce that $B(u,v)$ holds for any $u,v \in S$. But this is just the assertion (6.6) of the Theorem.

Further reading

D. Dewey, Microeconomics (Oxford, 1975), Chapters 13,14.
R.H. Coase. The problem of social cost. J. Law and Economics, 3 (1960), 1-44. (Entertaining as well as informative.)
Amartya K. Sen, Collective choice and social welfare, (Holden Day: Oliver and Boyd, 1970).

Chapter 4 Exercises

1. A family consists of m members. Subject to a budget constraint

$$\underline{p}\underline{x} \leq B$$

it purchases a bundle $\underline{x}$ of commodities and distributes it amongst its members so as to maximize the social utility

$$u(\underline{x}^1, \ldots, \underline{x}^m) \; .$$

(To each according to his needs!) Here $\underline{x}^i$ is the bundle received by member i of the family (so $\Sigma\, \underline{x}^i = \underline{x}$) and u is a function of the mn coefficients x_j^i (n = number of commodities) of the usual utility type. Show that the total amount x_1 of the first good is diminished if its price p_1 is increased (the prices of the other goods remaining constant).

[Hint. Chapter 1, Exercise 4.]

2. A factory produces wood pulp. The cost of producing x tonnes per annum is $f(x)$ pounds per annum. It also produces effluent which flows into a river and causes losses valued at $g(x)$ to the fisheries. Here $f(x), g(x)$ are increasing strictly convex functions of $x \geq 0$. The price at which the pulp may be sold is p per tonne (independent of the quantity x produced).

(i) Suppose that, by law, the factory owner must compensate the loss caused to the fisheries. Show that he will choose to produce y tonnes per annum, where y maximizes

$$px - f(x) - g(x) \; .$$

(ii) Suppose that a Free Enterprise government abrogates the law, and that no compensation need be paid. It is, however, open for the factory and fishery owners to strike a bargain that the latter will pay the former a sum s per annum on condition that the annual production of pulp is no more than t : here s,t are subject to negotiation. Show that they will agree on $s = g(y)$, $t = y$, and that a quantity y of pulp will be produced.

(iii) Suppose that a process is invented which renders the effluent innocuous and costs k per annum (independent of the amount of effluent). In case (i) show that the factory owner will install the process when $k < g(y)$ but not when $k > g(y)$. In case (ii) show that the fishery owner will install it under precisely the same circumstances.

[cf. R.H. Coase loc. cit.]

3. Consider the following model of whale fishery. In the absence of fishing, if there are y whales one year there are $y + f(y)$ the next, where $f(0) = 0$, $f(y)$ increases in $0 \le y \le y_1$, and decreases for $y \ge y_1$, passing through 0 at $y = y_2$. Show that the whale population in the absence of fishing tends to y_2 , and that the maximum number of whales which can be caught annually without ultimate extinction of the stock is $f(y_1)$.

A whaling boat costs $b > 0$ to operate for a year. It catches $g(y)$ whales annually, where $g(y)$ is a strictly increasing function of the stock y . Let p be the price of a whale, and define y^* by $g(y^*) = b/p$. Show that it is not profitable to engage in whaling if $y^* \ge y_2$. If $y^* < y_2$, show that in a competitive situation boats will enter the industry until the total number of boats is $N = f(y^*)/g(y^*)$. Show, however, if $y^* < y_1$ that the total catch would be improved if the total number of boats is restricted to a smaller number.

Because of technical improvements, the function $g(y)$ is replaced by $h(y) > g(y)$. Show that in a competitive situation the total catch will be reduced for certain values of the parameters.

4. (Lindahl prices.) Justify the unproved statements in the following account of a model which extends the notion of a (Walras) competitive allocation to include public goods. There are finitely many private goods and public goods. Bundles of private goods are denoted by small letters e.g. $\underline{x}$ and bundles of public goods by capitals e.g. $\underline{X}$.

The economy E consists of a finite number of households h . Each has a preference relation $\prec_h$, of the type introduced in Chapter 2, on bundles $(\underline{x},\underline{X})$ of combined private and public goods. Initially, there are no public goods and h has an endowment $\underline{w}_h$ of private goods.

There is a process which can create public goods from private goods. The set M of $(\underline{y},\underline{Y})$, where $\underline{Y}$ can be created from $\underline{y}$, is a closed convex cone with vertex $(\underline{o},\underline{O})$. If $(\underline{o},\underline{Y}) \in M$, then $\underline{Y} = \underline{O}$. (Public goods cannot be created from nothing.)

An allocation is given by a bundle $\underline{a}_h$ of private goods for each household, and a bundle $\underline{A}$ of public goods, subject to the condition that there is a bundle $\underline{a}^*$ of private goods with

$$\underline{a}^* + \Sigma\, \underline{a}_h = \Sigma\, \underline{w}_h$$

and

$$(\underline{a}^*,\underline{A}) \in M .$$

Show that the set of $\{\underline{a}_h\}$, $\underline{A}$ is bounded for given M and $\{\underline{w}_h\}$.

A Lindahl allocation is given by a price vector $\underline{p}$ for private goods together with price vectors $\underline{P}_h$ for public goods, one for each household. The allocation $\{\underline{a}_h\}$ is Lindahl for these prices if

(i) $(\underline{x},X) = (\underline{a}_h,\underline{A})$ maximizes h's utility subject to the budget constraint

$$\underline{px} + \underline{P}_h\underline{X} \le \underline{pw}_h \ . \qquad (*)$$

(ii) Either (α) $\underline{a}^* = \underline{o}$, $\underline{A} = \underline{O}$, or (β) $\underline{pa}^* = \underline{P}^*\underline{A}$, where $\underline{P}^* = \Sigma\, \underline{P}_h$ and $\underline{P}^*\underline{Y} \le \underline{py}$ for all $(\underline{y},\underline{Y}) \in M$.

Show that a Lindahl allocation is Pareto optimal.

A coalition $S \subset E$ is blocking for the allocation $\{\underline{a}_h\},\underline{A}$ if there is some other allocation for which the $h \notin S$ retain their initial endowments and which is preferred by all the $h \in S$ (strictly preferred by at least one of them). An allocation is in the core if there is no blocking coalition. Show that a Lindahl allocation is in the core.

The preceding does not require the indifference hypersurfaces (or, indeed, the cone M) to be convex. For the rest of the exercise we suppose that they are, but not necessarily strictly convex.

Let $\{\underline{a}_h\},\underline{A}$ be Pareto optimal. Show that there are prices $\underline{p},\{\underline{P}_h\}$ such that $\underline{a}_h,A$ maximizes h-utility subject to the budget constraint

$$\underline{px} + \underline{P}_h\underline{X} \le \underline{pa}_h + \underline{P}_h A \ .$$

(cf. Chapter 2, Exercise 7).

By modifying the argument of Chapter 2, §3 and its Addendum, show that a Lindahl allocation always exists. More precisely, let $\underline{p} >> \underline{O}$, $\underline{P}_h >> \underline{O}$ be prices. For each h let $(\underline{x},X) = (\underline{c}_h,\underline{C}_h)$ maximize h-utility subject to the budget constraint (*). Further, let $(\underline{d}^*,D)$ satisfy either (iiα) or (iiβ) above. Then there is an excess demand

$$\underline{e} = \underline{d}^* + \Sigma\, \underline{c}_h - \Sigma\, \underline{w}_h$$

for private goods. Further, household h has an excess demand

$$\underline{E}_h = \underline{C}_h - \underline{D}$$

for public goods. Here

$$\underline{pe} + \Sigma\, \underline{P}_h \underline{E}_h = 0 \ . \qquad (@)$$

For given $\underline{p}, \{\underline{P}_h\}$, the set of $\underline{e}, \{\underline{E}_h\}$ is convex. It depends upper semicontinuously on $\underline{p}, \{\underline{P}_h\}$ in the sense of the Addendum to Chapter 2, §3. Modify the definition of excess demand to cover the case when some of the prices are 0 . Then the generalization of Lemma 3.1 of Chapter 2 applies (with (@) the analogue of (3.14)).

[Cf. D.K. Foley. Lindahl's solution and the core of an economy with public goods. Econometrica 38 (1970), 66-72.]

CHAPTER 5. LINEAR ECONOMIC MODELS

1. Introduction

In this chapter we consider some very simple models of an economy in which production of goods from other goods can take place. There are a number of activities (or processes) P each of which can be run at any intensity $y \geq 0$. There are constant returns to scale, so that the input and output of yP are y times those of P. Further, two activities can be run simultaneously. If P_1, P_2 are two activities, then the input and output of P_1+P_2 are the sums of those for P_1, P_2. Sometimes we shall assume that there are a finite number $P_1,\ldots,P_m$ of basic activities such that every activity P is of the shape

$$P = \sum_i y_i P_i \ , \qquad (y_i \geq 0) \ . \qquad (1.1)$$

Here $\underline{y} = (y_1,\ldots,y_m) \geq 0$ is the intensity vector.

In general, goods can serve both as inputs and outputs to the activities. If a good is not part of the output of any activity, it is called a factor of production (or primary good).

Although prices do not occur in the formulation of the models, we shall see that they arise very naturally in their treatment.

2. Closed and open Leontieff models

We suppose, first, that there are n producible goods labelled $1,\ldots,n$ and no factors of production. There are also n basic activities all of which have the same period of operation (a "year"). Activity

P_j when operated at unit intensity produces a unit of good j at the end of the year and requires an input

$$\underline{a}(j) = (a_{1j},\ldots,a_{nj}) \geq \underline{0} \tag{2.1}$$

at the beginning of it. We regard vectors such as $\underline{a}(j)$, representing bundles of commodities, as column vectors (though we shall continue to write them horizontally to save trees). We combine the $\underline{a}(j)$ to make a square matrix

$$A = (a_{ij}) . \tag{2.2}$$

If the activities operate at intensity $\underline{y}$ (also a column vector), the input required at the beginning of the year is $A\underline{y}$ and the output at the end is $\underline{y}$. There is thus a <u>net product</u> of $\underline{y} - A\underline{y}$. We require that the net product is non-negative (e.g. because we are in a steady state), that is

$$\underline{y} - A\underline{y} \geq \underline{0} . \tag{2.3}$$

The system just described is sometimes called a <u>closed Leontieff model</u>.

<u>Theorem 2.1</u>. The following conditions are equivalent:

(i) for each j there is an intensity vector such that the net product includes a positive amount of good j ;

(ii) there is an intensity vector $\underline{y} > \underline{0}$ such that the net product contains a positive amount of every good: that is,

$$\underline{y} - A\underline{y} \gg \underline{0} ; \tag{2.4}$$

(iii) there is a price vector $\underline{p} > \underline{0}$ such that every activity makes a (strictly positive) profit: that is,

$$\underline{p} \gg \underline{p}A ; \tag{2.5}$$

(iv) every eigenvalue of A is strictly less than 1 .

Further, if these conditions are satisfied, for every bundle $\underline{x} > \underline{0}$ of commodities there is a unique intensity vector $\underline{y}$ such that $\underline{x}$ is the net product:

that is,

$$\underline{y} - A\underline{y} = \underline{x} \quad . \tag{2.6}$$

Proof. (ii) $\rightarrow$ (i) . Trivial.

(i) $\rightarrow$ (ii) . Take for the $\underline{y}$ in (ii) the sum of the intensity vectors in (i) for $j = 1,\ldots,n$.

(ii) $\leftrightarrow$ (iv) . This is an immediate consequence of the theory of non-negative matrices. More precisely, in the notation of Theorem 1 of Appendix C, both (ii) and (iv) are equivalent to $\mu(A) < 1$ (by (iv) and (ii) of the theorem respectively).

(iii) $\leftrightarrow$ (iv) . This is the dual of the preceding.

All the conditions are thus equivalent to $\mu(A) < 1$. If this holds we have

$$(I - A)^{-1} > 0 \tag{2.7}$$

by (v) of the theorem in the Appendix. Hence given $\underline{x} > \underline{0}$ the only solution of (2.6) is

$$\underline{y} = (I-A)^{-1}\underline{x}$$

$$> \underline{0} \quad . \tag{2.8}$$

This concludes the proof.

If we permit the introduction of factors of production we have an open Leontieff model. We suppose that there is a single factor of production ("labour") and that activity P_j at unit intensity requires

$$b_j > 0 \tag{2.8 bis}$$

units of labour (note the strict inequality). The vector $\underline{b} = (b_1,\ldots,b_n)$ is a row vector. For intensity $\underline{y}$ the labour requirement is $\underline{by}$.

Theorem 2.2. In the model just introduced, suppose that the price of a unit of labour ("wage") is 1 . The following statements are equivalent:

(i) there is an intensity vector $\underline{y}$ such that the net product contains every good in positive quantities, that is,

$$\underline{y} - A\underline{y} >> \underline{0} \quad . \tag{2.9}$$

(ii) there is a set of prices $\underline{p} = (p_1, \ldots, p_n)$ such that every activity at least breaks even, that is,

$$\underline{p} \geq \underline{p}A + \underline{b} \; . \qquad (2.10)$$

(iii) All eigenvalues of A are strictly less than 1 in absolute value.

If these conditions are satisfied, for every bundle $\underline{x} > \underline{0}$ of commodities there is a unique intensity vector $\underline{y}$ such that $\underline{x}$ is the net product. The amount of labour required is $\underline{vx}$, where

$$\underline{v} = \underline{b}(I - A)^{-1} \; . \qquad (2.11)$$

At prices $\underline{p} = \underline{v}$ every activity P_j precisely breaks even.

Proof. This is an almost immediate consequence of Theorem 2.1. Clearly condition (iii) of Theorem 2.1 is equivalent to (ii) of this theorem. Hence (i),(ii), (iii) are equivalent by Theorem 2.1. If $\underline{x},\underline{y}$ satisfy (2.6), the labour required is $\underline{by} = \underline{vx}$, where $\underline{v}$ is given by (2.11). Finally, $\underline{p} = \underline{v}$ gives equality in (2.10).

We note that (2.11) can be written

$$\underline{v} = \underline{b} + \underline{b}A + \ldots + \underline{b}A^k + \ldots , \qquad (2.12)$$

which has a heuristic interpretation. Let $\underline{x} > \underline{0}$ be a bundle. It is the gross output of working with intensity $\underline{x}$. This requires $\underline{bx}$ labour and an input of $A\underline{x}$. Now $A\underline{x}$ is the gross output of activity $A\underline{x}$, which requires $\underline{b}A\underline{x}$ labour and an input $A^2\underline{x}$. And so on. Hence the total labour input is

$$\underline{bx} + \underline{b}A\underline{x} + \underline{b}A^2\underline{x} + \ldots$$
$$= \underline{vx} \; . \qquad (2.13)$$

This confirms that $\underline{vx}$ is the amount of labour required to produce $\underline{x}$, and gives an interpretation of the individual summands in (2.12).

For later use we also note the

Corollary. Any set of prices $\underline{p}$ at which every activity breaks even (or better) satisfies $\underline{p} \geq \underline{v}$.

Proof. For (2.10) can be written

$$\underline{p}(I-A) \geq \underline{b} \,. \qquad (2.14)$$

The required result now follows from (2.7) and (2.11).

We now generalize. Instead of only one basic activity producing good j, we suppose that there may be several, say $P_j(m)$ $(1 \leq m \leq M_j)$, all of which may be run simultaneously at arbitrary (non-negative) intensities. Each such activity $P_j(m)$ (at unit intensity) requires an input $\underline{a}_j(m)$ (say) of goods and $b_j(m) > 0$ of labour and has an output of a single unit of good j. Again we consider only intensities

$$Y : \{y_j(m) \qquad (1 \leq m \leq M_j \,,\ 1 \leq j \leq n)\} \qquad (2.15)$$

such that the net product is non-negative.

A given bundle $\underline{x} > \underline{0}$ can now in general be obtained as a net product for many different sets of intensities (2.15). In general, the different Y will require different inputs of labour. We say that Y_0 yields $\underline{x}$ efficiently (as a net product) if the labour required is a minimum (over all Y yielding $\underline{x}$). It turns out that the multiplicity of alternative activities is no advantage. We can select n activities

$$P_j(m_j^*) \qquad (1 \leq j \leq n) \qquad (2.16)$$

such that every bundle $\underline{x}$ is yielded efficiently by a combination of them alone. More formally:

Theorem 2.3. For the system just described the two following statements are equivalent:

(i) some bundle $\underline{x} >> \underline{0}$ can be produced (as net product);

(ii) there are prices $\underline{p} > 0$ such that for each j at least one of the $P_j(m)$ does not make a loss (labour having unit price).

If these conditions are satisfied, then

(a) every bundle $\underline{x} > \underline{0}$ can be produced;

(b) there is a set $\underline{v} >> \underline{0}$ of prices such that none of the $P_j(m)$ makes a profit and that for each j

at least one, $P_j(m_j^*)$ (say), breaks even;

(c) every bundle $\underline{x} > \underline{O}$ can be produced efficiently by the use of a combination of the $P_j(m_j^*)$ alone. The amount of labour required to yield $\underline{x} > 0$ as net product is $\underline{vx}$.

Note. Both this theorem and its generalization to the Sraffa model (cf. Exercise 9) are called Samuelson's non-substitution theorem.

Proof. (ii) → (i). Immediate from Theorem (2.2).

(i) → (ii). Suppose that $\underline{x} >> \underline{O}$ is the net product of the activity

$$\sum_{j=1}^{n} \sum_{m=1}^{M_j} y_j(m) P_j \ . \qquad (2.17)$$

We can write this in the shape

$$\sum_{j=1}^{n} y_j' P_j' \ , \qquad (2.18)$$

where

$$P_j' = \sum_{m=1}^{M_j} u_j(m) P_j \qquad (2.19)$$

and the $u_j(m) \geq 0$ satisfy

$$\sum_m u_j(m) = 1 \quad (\text{all } j) \ . \qquad (2.20)$$

Then the P_j' have the properties postulated for the P_j in Theorem 2.2 (output at end of year is one unit of good j). Hence, by that theorem, (i) implies that there is a set of prices $\underline{p}$ such that each P_j' at least breaks even. But then, by (2.17) and linearity, for each j at least one of the $P_j(m)$ at least breaks even.

(a) Immediate from Theorem (2.2).

(b) The set S of $\underline{p}$ with the property (ii) is bounded away from zero (e.g. by Theorem 2.2 Corollary) and is closed. Hence Σp_j attains its minimum in S, say at $\underline{v}$. Suppose that one of the $P_j(m)$ makes a strictly positive profit at prices $\underline{v}$, say $P_k(m_1)$. Then it will continue to make a positive profit at prices $\underline{p}'$, where $p_k' = v_k - \delta$, $p_j' = v_j$ $(j \neq k)$, provided that $\delta > 0$ is small enough. Any $P_j(m)$ $(j \neq k)$ which breaks even or better at prices $\underline{v}$ will certainly do so at prices $\underline{p}'$. This contradicts the definition of $\underline{v}$. Hence no $P_j(m)$ makes a positive profit at $\underline{v}$. By the definition of S there is, however, for each j, a $P_j(m_j^*)$ which breaks even at $\underline{v}$.

(c) By Theorem 2.2 every $\underline{x} > \underline{0}$ can be obtained as the net product of a linear combination of the $P_j(m_j^*)$. Since the $P_j(m_j^*)$ break even at prices $\underline{v}$, the amount of labour required is $\underline{vx}$. On the other hand, no $P_j(m)$ makes a positive profit at $\underline{v}$. Hence for any combination (2.17) of activities which yields $\underline{x}$ as net product, the amount of labour required is at least $\underline{vx}$. Thus the production by the $P_j(m_j^*)$ is efficient.

Theorem 2.3 asserts the existence of prices $\underline{v}$ such that $\underline{vx}$ is the minimum amount of labour required to yield any given bundle $\underline{x} > \underline{0}$. Hence $\underline{v}$ can be regarded as a measure of the labour content. Such prices play an important role in the work of Ricardo and Marx and are described as *value* to distinguish them from other prices which may be considered. The existence of value depends, however, crucially on the rather strict conditions imposed. Theorem 2.2 breaks down if there is *joint production*, i.e. a basic process produces quantities of more than one good (as the sheep industry produces both wool and mutton). Again, there is no analogue of Theorem 2.3 when there is more than one factor of production (e.g. labour and land; or skilled and unskilled labour). [See Exercises.]

3. The Sraffa and Marx models

Throughout this section, we consider the situation to which Theorem 2.2 refers. There are n producible goods and one factor of production ("labour"). There are n activities P_j. Activity P_j requires an input of $\underline{a}(j)$ at the beginning of the year and also $b_j > 0$ units of labour: at the end of the year the output is a single unit of good j. The matrix A is given by (2.2) so that if the activities operate according to intensity vector $\underline{y}$ the input required is the bundle $A\underline{y}$ of goods and $\underline{by}$ labour; and the gross output is $\underline{y}$. We denote by $\underline{v}$ the value given by (2.9).

We suppose that the equivalent conditions (i),(ii), (iii) of Theorem 2.2 are satisfied, so that the largest positive eigenvalue $\mu(A)$ of A is <1 by (iii). We write

$$\{\mu(A)\}^{-1} = 1 + \Pi \tag{3.1}$$

where

$$\Pi > 0 \; . \tag{3.2}$$

Our own Piero Sraffa considered (Production of commodities by means of commodities) a system under which (i) labour is paid a wage $w > 0$, and (ii) the inputs are supplied by capitalists who charge a rate of interest π. Then the activities P_j no longer break even when evaluated at prices $\underline{v}$. However, prices $\underline{p}$ establish themselves by the usual economic forces under which all the P_j precisely break even. They must satisfy

$$(1+\pi)\underline{p}A + w\underline{b} = \underline{p} \; . \tag{3.3}$$

Since $\underline{b} \gg \underline{O}$ we must have

$$\mu(A)(1+\pi) < 1 \tag{3.4}$$

by the theory of non-negative matrices; that is

$$\pi < \Pi \tag{3.5}$$

by (3.1). Then (3.3) gives

$$w^{-1}\underline{p} = \underline{b} + (1+\pi)\underline{b}A + \ldots + (1+\pi)^k\underline{b}A + \ldots$$

$$> \underline{v} \qquad (3.6)$$

for $\pi > 0$, since $\pi = 0$ gives $w^{-1}\underline{p} = \underline{v}$ by (2.11). Hence the rapacity of capitalists has increased the level of prices, relative to the wage, above the value $\underline{v}$.

The system considered by Marx in Das Kapital is more elaborate than that of Sraffa. Each workman requires a certain minimum bundle $\underline{d}$ of commodities which is necessary for him to subsist for a year. The ruthless operation of capitalism ensures that he gets only $\underline{d}$ and no more. The value of the subsistence bundle $\underline{d}$ is the subsistence wage

$$s = \underline{vd} . \qquad (3.7)$$

On the other hand, the value of the labour provided by a workman is 1 by the definition of $\underline{v}$. Marx calls

$$1 - s \qquad (3.8)$$

the unpaid wage or surplus value. This is the part of the fruit of his labours of which the labourer is considered to be deprived. The ratio

$$\sigma = (1-s)/s \qquad (3.9)$$

is the rate of exploitation or rate of surplus value.

Marx supposes that the capitalists advance not only the goods required as inputs for the activities but also the goods required for the subsistence of the workforce. At activity $\underline{y}$ they therefore advance

$$A\underline{y} + (\underline{by})\underline{d} . \qquad (3.10)$$

The (gross) output is $\underline{y}$. As in the Sraffa model, Marx supposes that capitalists charge a rate of interest (or rate of profit) π and that goods are exchanged at prices $\underline{p}$ for which each activity just breaks even. Hence π and $\underline{p}$ satisfy

$$(1+\pi)\underline{p}(A+\underline{db}) = \underline{p} . \qquad (3.11)$$

Here $\underline{db}$, the product of the column vector $\underline{d}$ and the row vector $\underline{b}$, is a square $n\times n$ matrix of rank 1. The theory of positive matrices (Appendix C) shows that

$(1+\pi)^{-1}$ is the largest positive eigenvalue of $A+\underline{db}$. [If $A+\underline{db}$ is irreducible, this follows from the Corollary to the Theorem in Appendix C. If this condition is not satisfied, any possible other solutions $\pi, \underline{p}$ of (3.11) are dismissed as being economically meaningless.]

<u>Theorem 3.1</u>. $\pi > 0$ precisely when $\sigma > 0$, and then

$$\pi \leq \sigma . \qquad (3.12)$$

<u>Proof</u>. By (3.11) we have

$$\mu(A+\underline{db}) = (1+\pi)^{-1} , \qquad (3.13)$$

and so there is a bundle of goods $\underline{e} > \underline{O}$ such that

$$(1+\pi)(A+\underline{db})\underline{e} = \underline{e} . \qquad (3.14)$$

Further, by (2.9), (3.7), (3.9) we have

$$\underline{v}\{A + (1+\sigma)\underline{db}\} = \underline{v} . \qquad (3.15)$$

On acting with $\underline{v}$ on (3.14) on the left and with $\underline{e}$ on (3.15) on the right, we obtain

$$\sigma(\underline{vd})(\underline{be}) = \pi\{\underline{vAe} + (\underline{vd})(\underline{be})\} . \qquad (3.16)$$

Here the scalar products $\underline{vd}$, $\underline{be}$ are strictly positive because we have supposed by (2.8 bis) that $\underline{b} \gg \underline{O}$, so $\underline{v} \gg \underline{O}$ by (2.11). Further, $\underline{vAe} \geq 0$. Hence (3.16) implies all the statements of the Theorem.

Marx regarded the distinction between $\underline{v}$ and $\underline{p}$ as fundamental to economics (the "transformation of values into prices of production"). However, textual critics say that he did in fact muddle them. Further, his metaphysical predilections led him to assert relations which are in fact not true as mundane propositions of linear algebra. For all this, we refer to the books of Morishima and Pasinetti listed at the end of the chapter.

4. <u>The Gale economy</u>

Here there are no factors of production: every good may be produced by some activity. It is thus assumed either that sufficient labour is available so as not to be a limitation or, alternatively, that it is produced from inputs of food, clothing etc. Joint production is permitted and, indeed, is regarded as the rule.

We suppose that all activities have the same period of operation (a "year"). We first look at the economy over a single year, and will subsequently (§6) consider a sequence of years. There are n goods. We shall denote an activity by the pair $(\underline{u},\underline{v})$ where $\underline{u} \geq \underline{O}$ is the input and $\underline{v} \geq \underline{O}$ the output [if two activities have the same $\underline{u}$ and $\underline{v}$ they are identified]. The set of $(\underline{u},\underline{v})$ is the economy E, which we regard as a subset of R^{2n}. We suppose that E is a convex cone, that is, that

(i) if $(\underline{u},\underline{v}) \in E$ then $(\lambda\underline{u}, \lambda\underline{v}) \in E$ for all $\lambda > 0$;

(ii) if $(\underline{u}^1,\underline{v}^1),(\underline{u}^2,\underline{v}^2) \in E$ then $(\underline{u}^1+\underline{u}^2,\underline{v}^1+\underline{v}^2) \in E$.

We also make the following assumptions.

Assumption 1. E is a closed subset of R^{2n}.

This is primarily a technical convenience.

Assumption 2. If $(\underline{O},\underline{v}) \in E$, then $\underline{v} = \underline{O}$. (4.1)

(No free lunch assumption.)

Assumption 3. Every good occurs in the output of some activity (i.e. given j there is some $(\underline{u},\underline{v}) \in E$ with $v_j > 0$).

For, otherwise, the good would disappear from the economy after the first year.

We now define the technological expansion rate α to be the supremum of the $a \geq 0$ such that $\underline{v} \geq a\underline{u}$ for some $(\underline{u},\underline{v}) \in E$ with $\underline{v} \neq \underline{O}$.

Lemma 4.1. $0 < \alpha < \infty$. (4.2)

There is an $(\underline{u}^*,\underline{v}^*) \in E$ such that

$$\underline{v}^* \geq \alpha\underline{u}^* > \underline{O}. \tag{4.3}$$

Proof. By homogeneity we may restrict attention to $(\underline{u},\underline{v})$ in

$$\Sigma u_j + \Sigma v_j = 1. \tag{4.4}$$

By Assumption 3, there is $(\underline{u},\underline{v})$ with $\underline{v} \gg \underline{O}$, so $\alpha > 0$. If $\alpha = \infty$, so a can be arbitrarily large, then by Assumption 1 and the compactness of (4.4) we should have a $(\underline{u},\underline{v})$ on (4.4) with $\underline{u} = \underline{O}$, contrary to Assumption 2. Finally, the existence of $(\underline{u}^*,\underline{v}^*)$ satisfying (4.3) again follows from a compactness argument.

The <u>economic expansion rate</u> β is defined to be the infimum of the b for which there exists a $\underline{p} > 0$ with $\underline{pv} \le b\underline{pu}$ for all $(\underline{u},\underline{v}) \in E$.

<u>Lemma 4.2</u>. $0 < \beta$. (4.5)

If $\beta \neq \infty$ (which will be proved in Lemma 4.3), then there is a $\underline{p}^* > \underline{O}$ such that

$$\underline{p}^*\underline{v} \le \beta\underline{p}^*\underline{u} \quad (4.6)$$

for all $(\underline{u},\underline{v}) \in E$.

<u>Proof</u>. We need consider only $\underline{p}$ in the compact set

$$\Sigma \underline{p}_j = 1 . \quad (4.7)$$

If $\beta \neq \infty$, then the obvious compactness argument gives a $\underline{p}^*$ satisfying (4.6). If $\beta = 0$, (4.6) is a contradiction to Assumption 3, so (4.5) holds.

<u>Lemma 4.3</u>.

$$\beta \le \alpha . \quad (4.8)$$

<u>Proof</u>. The cone $W \subset R^n$ consisting of the

$$\underline{w} = \underline{v} - \alpha\underline{u} \qquad (\underline{u},\underline{v}) \in E \quad (4.9)$$

is convex, since E is convex. It is disjoint from

$$\Omega : \underline{w} \gg \underline{O} : \quad (4.10)$$

for if $\underline{v}^o - \alpha\underline{u}^o \gg \underline{O}$ for some $(\underline{u}^o,\underline{v}^o) \in E$, then $\underline{v}^o \ge a\underline{u}^o$ for some $a > \alpha$, contradicting the maximality of α. By a standard result on convex cones (Lemma 8 of Appendix A), there is a $\underline{p} > \underline{O}$ such that

$$\underline{pw} \le 0 \qquad (\text{all } \underline{w} \in W) ; \quad (4.11)$$

that is

$$\underline{pv} \le \alpha\underline{pu} \qquad (\text{all } \underline{u},\underline{v}) \in E) . \quad (4.12)$$

Hence we can take $b = \alpha$ in the definition of β, and (4.8) follows.

We say that the economy E is <u>reducible</u> if there is a non-empty proper subset T of the goods $1,\ldots,n$ which can be produced using only the goods in T. (In other words, if E_T consists only of the $(\underline{u},\underline{v}) \in E$ for which the bundles $\underline{u},\underline{v}$ contain only goods in T, then E_T is a Gale economy on T satisfying Assumption 3.) If E is not reducible, it is <u>irreducible</u>.

<u>Lemma 4.4</u>. Suppose that E is irreducible. Then

$$\underline{v}^* >> \underline{0} \,, \tag{4.13}$$

$$\beta = \alpha \tag{4.14}$$

and

$$\underline{p}^*\underline{v} = \alpha \underline{p}^* u^* \,. \tag{4.15}$$

<u>Proof</u>. Denote by T the set of t for which $v_t^* > 0$. Then $u_j^* = 0$ for $j \notin T$ by (4.3). If T is a proper set of the goods, then it has all the properties required in the definition of reducibility. Hence irreducibility implies (4.13).

By (4.3) and (4.6) we have

$$\alpha \underline{p}^* \underline{u}^* \le \underline{p}^* \underline{v}^* \le \beta \underline{p}^* \underline{u}^* \,, \tag{4.16}$$

where $\underline{p}^*\underline{v}^* > 0$ by (4.13). Hence $\alpha \le \beta$, so (4.14) holds by Lemma 4.3. Finally, (4.15) follows from (4.14) and (4.16).

<u>Corollary</u>. Let j be a good for which $v_j^* > \alpha u_j^*$. Then j is free in $\underline{p}^*$.

For we can write (4.15) in the shape

$$\underline{p}^*(\underline{v}^* - \alpha \underline{u}^*) = 0 \tag{4.17}$$

and invoke (4.3).

5. <u>von Neumann model</u>

This is the special case when every activity P is a combination

$$P = \sum_i y_i P_i \qquad y_i \ge 0 \tag{5.1}$$

of a finite number m of basic activities

$$P_i = (\underline{u}^i, \underline{v}^i) \qquad (1 \le i \le m) \,. \tag{5.2}$$

Here $\underline{y} = (y_1,\ldots,y_m) \ge \underline{0}$ is the intensity vector.

Assumption 1 of §4 is automatic. Assumption 2 is satisfied if

$$\underline{u}^i \neq \underline{0} \qquad (1 \le i \le m) . \qquad (5.3)$$

Finally, Assumption 3 holds provided that for each good j there is some i such that $v_j^i \neq 0$.

It is left to the reader to translate the results of §4 into this new framework and we mention only

Lemma 5.1. Suppose that the economy is irreducible, and let the intensity vector $\underline{y}^*$ correspond to $(\underline{u}^*, \underline{v}^*)$. Suppose that activity P_i does not give a rate of return α at prices $\underline{p}^*$, that is that

$$\underline{p}^* \underline{v}^i < \alpha \underline{p}^* \underline{u}^i . \qquad (5.4)$$

Then $y_i^* = 0$ (i.e. the activity is not indulged in).

Proof. Equation (4.16) becomes

$$\sum_i y_i (\alpha \underline{p}^* \underline{u}^i - \underline{p}^* \underline{v}^i) = 0 , \qquad (5.5)$$

where all the summands are non-negative by (4.6) and (4.15).

6. Turnpike theorems

We revert to the discussion of an irreducible Gale economy E in §4, and retain the same notation. Hitherto we have looked at it over a single "year". We now consider it over a succession of years, and will suppose that

$$(\beta =) \quad \alpha > 1 , \qquad (6.1)$$

so that the stock of goods in the economy can expand. Each year t starts with a stock $\underline{s}^{t-1}$ of goods. An activity is chosen of which this is the input. The output is the stock for next year. We thus get a sequence $\underline{s}^o, \underline{s}^1, \underline{s}^2, \ldots$ of stocks such that

$$(\underline{s}^{t-1}, \underline{s}^t) \in E \qquad (t = 1,2,\ldots) . \qquad (6.2)$$

Given the initial stock, the object of good economic management is to choose a sequence of activities so that the sequence $\underline{s}^o, \underline{s}^1, \ldots$ is optimal in some way to be decided.

A limitation to the growth of the sequence $\underline{s}^t$ is given by (4.6), which (with (4.15)) implies that

$$\underline{p}^*\underline{s}^t \leq \alpha^t \underline{p}^*\underline{s}^o . \qquad (6.3)$$

To obtain results in the opposite direction, it is convenient to assume that there is free disposal, that is that at any time the economy can diminish its stock of goods without penalty. In symbols this is the assumption that

$$\underline{u}^o \geq \underline{u} , \ \underline{v} \geq \underline{v}^o, \ (\underline{u},\underline{v}) \in E \Rightarrow (\underline{u}^o,\underline{v}^o) \in E . \qquad (6.4)$$

With free disposals, (4.3) implies that

$$(\underline{v}^*, \alpha\underline{v}^*) \in E . \qquad (6.5)$$

On taking $\underline{s}^o = \underline{v}^*$, we see that (6.3) is best possible.

More generally, if we assume that $\underline{s}^o \gg \underline{0}$, then there is some $\lambda > 0$ such that

$$\lambda\underline{v}^* \leq \underline{s}^o \qquad (6.6)$$

and we may choose the sequence

$$\underline{s}^t = \lambda\alpha^{t-1}\underline{v}^* \qquad (t > 0) \qquad (6.7)$$

for which $\underline{p}^*\underline{s}^t$ differs from the right-hand side of (6.3) only by a constant independent of t .

So far we have been valuing the stocks at von Neumann prices $\underline{p}^*$. However, if $\underline{p}^* \gg \underline{0}$ (there are no free goods at von Neumann prices) and if $\underline{p} \gg \underline{0}$ is another price vector, then $\underline{p}^*\underline{x}/\underline{p}\underline{x}$ lies between positive constants for $\underline{x} > \underline{0}$. Hence for the asymptotic behaviour of $\underline{s}^t$ it makes little difference whether we use $\underline{p}$ or $\underline{p}^*$.

It will be observed that all of the stocks (6.7) lie on the ray

$$R = \{\lambda\underline{v}^* : \lambda > 0\} . \qquad (6.8)$$

There are a number of theorems which state that, under appropriate conditions, most of the stocks $\underline{s}^t$ in a sequence which is optimal in some sort of way will lie not far from R . Such theorems are known as turnpike theorems from an American word meaning "trunk road". The optimal way to develop the economy is thus (in this model) largely independent of the original stock and of the ultimate goal: if you wish large quantities of candy-floss, you should promote heavy industry!

We shall now enunciate and prove a simple turnpike theorem. By a <u>conical neighbourhood</u> of R we shall mean an open cone which contains R .

<u>Theorem 6.1</u>. Suppose that $\underline{p}^* \gg \underline{O}$, that any $(\underline{u},\underline{v}) \in E$ such that

$$\underline{p}^*\underline{v} \geq \alpha\underline{p}^*\underline{u} \qquad (6.9)$$

is a multiple of $(\underline{v}^*,\alpha\underline{v}^*)$, and that there are free disposals. Let prices $\underline{p} \gg \underline{O}$, an initial stock $\underline{s}^O \gg \underline{O}$ and a conical neighbourhood N of the turnpike R be given. Then there is a constant K , depending only on the economy E and on $\underline{p},\underline{s}^O,N$, with the following property:

Let T be given and choose the sequence of stocks

$$\underline{s}^O, \underline{s}^1, \ldots, \underline{s}^T \qquad (6.10)$$

so as to maximize $\underline{ps}^T$. Then at most K of the $\underline{s}^t$ $(O \leq t \leq T)$ lie outside N .

<u>Note</u>. For an economy of the type discussed in §5, the condition involving (6.9) can be satisfied only if $(\underline{v}^*,\alpha\underline{v}^*)$ is a multiple of a basic activity, as is easily verified, so the theorem is really only of interest for more general Gale economies.

<u>Proof</u>. The sequence (6.7) has $\underline{ps}^T = C_O\alpha^T$, where C_O depends only on $\underline{p},\underline{s}^O$ (and the economy E). Hence for a sequence maximizing $\underline{ps}^T$ we have

$$\underline{p}^*\underline{s}^T \geq C_1\alpha^T \qquad (6.11)$$

for some constant C_1 .

By the first hypothesis of the enunciation, there is a $\delta > O$ such that

$$\underline{p}^*\underline{v} \leq (\alpha-\delta)\underline{p}^*\underline{u} \qquad (6.12)$$

for $(\underline{u},\underline{v}) \in E$ unless $\underline{u},\underline{v}$ both lie in the conical neighbourhood N . Hence

$$\underline{p}^*\underline{s}^t \leq (\alpha-\delta)\underline{p}^*\underline{s}^{t-1} \qquad (6.13)$$

unless $\underline{s}^t$ and $\underline{s}^{t-1}$ lie in N . If L is the number of t $(O \leq t \leq T)$ for which $\underline{s}^t \notin N$, we thus have

$$\underline{p}^* \underline{s}^T \leq (\alpha-\delta)^L \alpha^{T-L} \underline{p}^* \underline{s}^o . \tag{6.14}$$

On comparing (6.11) and (6.14) we have

$$\theta^L \geq C_2 > 0 ,$$

where

$$\theta = (\alpha-\delta)/\alpha < 1$$

and C_2 is a constant. Hence L is bounded, as asserted.

Further reading

David Gale, The theory of linear economic models (McGraw-Hill, 1960).

For §3:

P. Sraffa, Production of commodities by means of commodities (Cambridge U.P., 1960).

M. Morishima, Marx's economics (Cambridge U.P., 1973).

L.L. Pasinetti, Lectures on the theory of production (Columbia U.P., 1977).

For §6:

D. Gale, The closed linear model of production. In Linear inequalities and related systems, Annals of Math. Studies 38, Princeton U.P., 1956.

Chapter 5 Exercises

1. In the model considered in Theorem 2.2 suppose that technical progress reduces the amount of labour required in activity P_k to $b_k' < b_k$. Show that the new value vector $\underline{v}'$ satisfies $\underline{v}' < \underline{v}$. Show, further, that $v_j'/v_j \geq v_k'/v_k$ for every j.

   Similarly, consider the effect of diminishing the amount a_{ik} of good i required as an input to P_k.

   [Hint. For $j \neq k$ one has both $\underline{va}(j) + b_j = v_j$ and $\underline{v}'\underline{a}(j) + b_j = v_j'$. Consider the j for which v_j'/v_j is minimal.]

   [Morishima: Marx's economics.]

2. The activity P_1 requires an input of ½ unit of good 1 and produces one unit of good 2 and one unit of good 3. Processes P_2, P_3 are similar with cyclic permutation of the indices 1,2,3. Each process P_j requires one unit of labour. Find the bundles $\underline{x} \geq \underline{0}$ that can be the net product of an intensity vector $\underline{y}$ requiring one unit of labour. Show that the conclusion of Theorem 2.2 fails to hold.

   [Gale: Linear economic models.]

3. Suppose that there are two factors of production (skilled and unskilled labour) and two produced goods. Activity P_1 produces one unit of good 1 using one unit of skilled labour. Activity P_2 produces one unit of good 2 using one unit of skilled labour and activity P_3 produces one unit of good 2 using two units of unskilled labour. One unit each of skilled and unskilled labour is available. Show that the set of outputs that can be efficiently produced by combinations of any two activities is strictly smaller than the set that

can be produced by using all three. (Contrast Theorem 2.3.)

[Gale.]

4. The open Leontieff model of Theorem 2.2. is said to have _uniform capital intensity_ (or _uniform organic composition of capital_) if $\underline{b}$ is an eigenvector of A for eigenvalue $\mu(A)$.

Consider a Sraffa Model with uniform capital intensity and fixed rate of interest π. Show that one can take $\underline{p} = \underline{v}$ and that then the wage w is given by

$$\pi = \Pi(1-w) .$$

[Pasinetti: _Lectures on theory of production_.]

5. In the open Leontieff model of Theorem 2.2 show that there is a bundle $\underline{s} > \underline{0}$ which is an eigenvector of A for eigenvalue $\mu(A)$. When it is normalized by $\underline{vs} = 1$ it is _Sraffa's standard net product_. Show that

$$\underline{bs} = \{\Pi/(1+\Pi)\}$$

in the notation of §3.

In the Sraffa model let $\underline{p}$ be normalized by $\underline{ps} = 1$. Show that

$$\pi = \Pi(1-w) .$$

[Pasinetti. _loc.cit_.]

6. Consider Marx's model when there is uniform intensity of capital [Exercise 4]. In the notation of §3 show that $\pi = \Pi(1-\delta)/(1+\delta\Pi)$, where $\delta = \underline{vd}$, and confirm Theorem 3.1.

[Pasinetti. _loc cit_.]

7. Consider Marx's model when the subsistence bundle $\underline{d}$ is a multiple $\delta\underline{s}$ of Sraffa's standard net product [Exercise 5]. Show that π is given by the same formula as in the preceding exercise, and confirm Theorem 3.1.

[Pasinetti. _loc cit_.]

8. The manufacture of a certain good requires the input only of labour, but there are two alternative processes. To have a unit of the good now, alternative P_1 requires 27 units of labour a year ago. Alternative P_2, however, requires 10 units of labour two years ago together with 18 units now. Let π be the rate of profit (in the sense of §3). Show that P_2 will be chosen if $0.2 < \pi < 0.5$ but P_1 if either $\pi < 0.2$ or $\pi > 0.5$.

   [Note. This examplifies the phenomenon of reswitching, which is regarded by some economists as paradoxical.]

9. (Reswitching in Sraffa model.) This exercise extends the simple Sraffa model discussed at the beginning of §2 to the situation discussed in Theorem 2.3, when there are several activities $P_j(m)$ $(1 \le m \le M_j)$ producing good j $(1 \le j \le n)$.

   (i) For fixed rate of profit π, prove an analogue of Theorem 2.3. Denote by $\underline{p}(\pi)$ the price vector corresponding in this analogue to $\underline{v}$ (normalized by $w = 1$): and let the set of activities corresponding to the $P_j(m_j^*)$ be

   $$P_j(m_j(\pi)) \qquad (1 \le j \le n) . \qquad (@)$$

   (ii) If π is now allowed to vary, show that in general $m_j(\pi)$ depends on π.

   (iii) Show that $\underline{p}(\pi_1) \le \underline{p}(\pi_2)$ whenever $\pi_1 \le \pi_2$.

   (iv) Suppose that we are interested only in the production of good 1 and regard goods $2,\ldots,n$ as "intermediates". Let $L(\pi)$ be the amount of labour required to produce a single unit of good 1 (as net product) by the set (@) of processes which are adopted when the rate of interest is π. Show that normally $L(\pi)$ is a stepwise constant function, not well-defined at points of discontinuity. (In exceptional cases there can be intervals in which $L(\pi)$ is not well-defined). Give an example to show that $L(\pi)$ need not be a monotone function of π.

[Hint. Previous exercise. Note. The result that $L(\pi)$ may decrease when π increases is contrary to some intuitions about the effect of interest rates on labour-intensity of production. For discussion, cf. Burmeister, Capital theory and dynamics, Chapter 4.]

10. (a) Let E be an irreducible Gale economy. Let $\gamma > 0$ be real and suppose that

(i) there is a price vector $\underline{p}^o > 0$ such that $\underline{p}^o\underline{v} \leq \gamma\underline{p}^o\underline{u}$ (all $(\underline{u},\underline{v})$).

(ii) There is a $(\underline{u}^o,\underline{v}^o) \in E$ such that $\underline{v}^o \geq \gamma\underline{u}^o > \underline{0}$.

Show that $\alpha = \beta = \gamma$.

(b) Show that the von Neumann economy E with 4 goods and the 3 basic activities $(\underline{u}^j,\underline{v}^j)$ $(1 \leq j \leq 3)$:

$$\underline{u}_1 = (0,1,0,0)\ ,\ \underline{v}_1 = (1,0,0,0)$$
$$\underline{u}_2 = (1,0,0,1)\ ,\ \underline{v}_2 = (0,0,2,0)$$
$$\underline{u}_3 = (0,0,1,0)\ ,\ \underline{v}_3 = (0,1,0,1)$$

is irreducible. By considering the intensity vector $\underline{y}^o = (\delta,\delta^2,1)$ and the price vector $\underline{p}^o = (1,\delta,\delta^2,0)$, where $\delta^3 = 2$, show that $\alpha = \delta^{-1}$.

[Gale.]

11. Let E be a Gale economy and let U be a bounded set in R^n. Show that the set of $\underline{v}$ for which there is a $\underline{u} \in U$ with $(\underline{u},\underline{v}) \in E$ is also bounded.

12. Let E be a Gale economy, and let E^o be the set of $(\underline{u}^o,\underline{v}^o) \in R^{2n}$ for which there is some

$(\underline{u},\underline{v}) \in E$ with $\underline{u} \leq \underline{u}^o$, $\underline{v} \geq \underline{v}^o$. Show that E^o is also a Gale economy. Show that E^o has free disposal, and that it has the same technological expansion rate and economic expansion rate as E.

13. Let $a > b > 1$. Consider the set E of pairs of vectors $(\underline{u},\underline{v})$ with $\underline{u},\underline{v} \in R^2$ and

$$0 \leq v_1 \leq au_2 \,,\quad 0 \leq v_2 \leq bu_1 \,.$$

Show that E is an irreducible Gale economy, and determine the expansion rate α.

Let T be a positive integer and $\underline{s}^o = (1,1)$. Determine the sequence of stocks $\underline{s}^t$ with $(\underline{s}^{t-1},\underline{s}^t) \in E$ which maximizes $\underline{ps}^{\overline{T}}$ for $\underline{p} = (1,1)$. Show that the conclusion of the Turnpike Theorem does not hold.

14. Show that the results of §6 continue to hold if free carry-over is supposed instead of free disposal. Free carry-over is defined by

$$\underline{x} \geq \underline{0} \,,\ (\underline{u},\underline{v}) \in E \Rightarrow (\underline{u}+\underline{x},\underline{v}+\underline{x}) \in E \,.$$

15. ("catenary property"). Under the conditions of Theorem 6.1, show that there is an L^* (depending only on $\underline{p},\underline{s}^o,N$) such that $\underline{s}^t \in N$ for all t in $L^* \leq t \leq T-L^*$.

[Hint. Normalize $\underline{v}^*$ so $\underline{p}^*\underline{v}^* = 1$. Let $\eta > 0$ be small. There is a conical neighbourhood N^* of the turnpike R such that for every $\underline{x} \in N^*$ we have

$$(1-\eta)(\underline{p}^*\underline{x}^*)\underline{v}^* < \underline{x} < (1+\eta)(\underline{p}^*\underline{x}^*)\underline{v}^* \,.$$

Let L^* be the value of L corresponding to N^*. Then there are $t(1) \leq L^*$, $t(2) \geq T-L^*$ such that $\underline{s}^{t(1)} \in N^*$, $\underline{s}^{t(2)} \in N^*$. We can now consider replacing $\underline{s}^t$ by (i) freely disposing at $t(1)$ to get on the turnpike, staying on the turnpike until $t(2)$ and then freely disposing to get a multiple $\rho\underline{s}^{t(2)}$ of $\underline{s}^{t(2)}$, (ii) replacing $\underline{s}^t$ by $\rho\underline{s}^t$ for $t \geq t(2)$. If $\rho > 1$, this contradicts optimality, so $\rho \le 1$. Hence

$$\underline{p}^*\underline{s}^{t(2)}/\underline{p}^*\underline{s}^{t(1)} > (1-\eta)(1+\eta)^{-1}\alpha^{t(2)-t(1)}$$

$$> (\alpha-\delta)\ \alpha^{t(2)-t(1)-1},$$

if η is chosen appropriately, and the result follows.]

CHAPTER 6. SIMPLE MACROECONOMIC MODELS

1. Introduction

Macroeconomic theory considers the working of the economy as a whole, and deals in large aggregates such as "total income", "total demand", "saving", "investment" and the like. It is not always clear exactly what these terms mean, even less how they should be measured in any given situation; and different economists have taken different interpretations. We shall adopt the eminently respectable tradition of pushing these difficulties to one side.

One must consider the ways in which these large aggregates affect each other. It is almost certainly true that everything affects everything else. We shall single out the influences we treat as important, and ignore the others. Economists of different epochs or of different schools have considered different influences to be the important ones, and so arrived at radically different theories: there is less concensus in macroeconomics than in microeconomics. We shall follow the paradigm of Samuelson's Economics, itself based on the ideas of Keynes. We start from simple models and work up to more sophisticated ones, sometimes modifying (or even abandoning) hypotheses made earlier.

Some classical economists have asserted that "money is a veil", i.e. that it obscures our view, but does not affect the workings, of the "real economy". Hitherto we have indeed treated prices only as constructs which facilitate the mathematical treatment of the "real economy". We shall now have to consider the interaction of the monetary with the "real" economy.

The economy we consider will be "closed"; that is, there is no external trade. We shall be concerned only with a static situation, though we may compare static situations with different values of the parameters (comparative statics). We shall even ignore questions of stability, as their discussion requires "economic dynamics", at least in a rudimentary form.

A list of the many symbols which are necessarily introduced is given for reference at the end of the Chapter.

2. An ultrasimple model

Consider the three quantities

(i) Aggregate demand, consisting of consumers' purchases, investment purchases, government purchases etc. We denote by Z the value at current prices.

(ii) The output Q of goods and services produced by businesses etc. This also is valued at current prices.

(iii) The total income Y of the factors of production, that is, wages, rents, etc.

There is a cyclical relationship

$$Z \to Q \to Y \to Z \,. \tag{2.1}$$

Aggregate demand Z calls forth output Q. The proceeds of the sale of output accrue as income Y to the factors of production. Finally, the spending of the income Y gives rise to the demand Z. If the economy is out of equilibrium, there will, in general, be lags in the cycle (2.1) as the system evolves in time. We are concerned here only with equilibrium conditions, and then

$$Z = Q = Y \,. \tag{2.2}$$

Demand falls into two parts:

$$Z = C + I \,. \tag{2.3}$$

Here C is the demand for consumption and I is the demand for investment goods (which add to the country's capital stock, or replace that which has become obsolete).

Income Y also falls into two parts:

$$Y = C + S . \tag{2.4}$$

Here C is the spending on consumption and S is saving.

On combining (2.2), (2.3) and (2.4) we have

$$S = I . \tag{2.5}$$

For the present, I is regarded as fixed by the nature of the economy. On the other hand, consumption and saving are taken to be increasing functions $C(Y)$, $S(Y)$ of income, where, of course,

$$C(Y) + S(Y) = Y . \tag{2.6}$$

We call the derivatives $c = C'(X)$, $s = S'(Y)$ the marginal propensity to consume and to save respectively and suppose that

$$0 < c < 1 , 0 < s < 1 , \tag{2.7}$$

where

$$c + s = 1 \tag{2.8}$$

by (2.6).

The value $Y = Y_o$ for equilibrium is determined by (2.5), that is

$$S(Y_o) = I . \tag{2.9}$$

We now consider what happens if we change the specification of the economy. Suppose, first, that investment is increased by a (small) quantity δ. Then by (2.9) the new value Y_1 of Y is

$$Y_1 \doteqdot Y_o + \delta/s . \tag{2.10}$$

Here $1/s > 1$ is Keynes' famous multiplier. (Other "multipliers" occur in the theory.)

Now suppose instead that I remains unchanged but that people become more thrifty. Then the function $S(Y)$ is replaced by $S^*(Y)$, where

$$S^*(Y) > S(Y) \qquad \text{(all } Y) . \tag{2.11}$$

The equilibrium value Y^* of Y is now given by

$$S^*(Y^*) = I . \tag{2.12}$$

Clearly

$$Y^* < Y_o \ . \qquad (2.13)$$

Hence total income is diminished and the amount saved

$$S(Y_o) = S^*(Y^*) = I \qquad (2.14)$$

remains the same. This is the Paradox of Thrift. If, instead of treating investment I as a constant, we had assumed, not implausibly, that it is a slowly increasing function $I(Y)$ of Y, then the paradox would have been sharpened: increased thrift leads to an actual decrease in the amount saved! [But cf. end of section 6.]

Note. In what follows, we shall generally denote the quantity (2.2) by Y and refer to it as the GNP (gross national product), though this is strictly Q.

3. Government

We now introduce government into the model. This has two effects.

(i) There is a demand G from government for goods and services. Hence total demand must now be written

$$Z = C + I + G \ , \qquad (3.1)$$

where C, I (as before) are (private) consumption and (private) investment.

(ii) Government levies a tax T. It is natural to assume that consumption and savings are functions of income after tax:

$$C = C(Y-T) \ , \ S = S(Y-T) \ . \qquad (3.2)$$

Since $Y = Z$ in equilibrium, we now have

$$Y = C(Y-T) + I + G \qquad (3.3)$$

or, what is the same,

$$S(Y-T) = I + G - T \ . \qquad (3.4)$$

If we suppose that G, T (and I) are fixed, then (3.4) determines the gross national product Y.

If G is increased by δ, the effect on Y is the same as an increase of δ in I, that is Y is increased by δ/s approximately, where $1/s$ is the multiplier.

If T is increased by δ, then Y is decreased

by $\delta(1-s)/s$ approximately. Under the reasonable assumption that $s < 0.5$, we have again a multiplier > 1.

If both G and T are increased by δ, then Y is increased by δ (exactly).

Now, instead of considering taxation T as fixed, suppose that it is an increasing function $T(Y)$ of Y. It is readily verified that the multiplier for changes in G or I is now smaller than before, namely

$$1/\{s + \tau(1-s)\} \qquad (3.5)$$

where $\tau = T'$.

Finally, suppose also that G is a function $G(Y)$ of Y. The natural assumption to make here is that G decreases when Y increases. When Y is small, industrial activity is low and so the State must pay unemployment benefit etc. Put $\gamma = -G' > 0$. Then the multiplier for I is yet further decreased, namely

$$1/\{s + \tau(1-s) + \gamma\}. \qquad (3.6)$$

4. Employment

Hitherto we have adopted the classical assumption that the operation of the markets equates supply and demand. Our own Professor Pigou wrote "With perfectly free competition among work-people ... everyone will be employed" (Theory of unemployment) - this in 1933 when unemployment in the U.K. was about 20%.

As a simple way of introducing the possibility of unemployment into the model, we assume that there is a critical value Y_F of the GNP (the full-employment GNP) such that

(i) if $Y < Y_F$, there is unemployment;

(ii) $Y > Y_F$ cannot occur, at least in an equilibrium state. (An attempt to achieve it leads to "overheating".)

We recall that the equilibrium GNP Y_O is given by

$$Y_o = C(Y_o - T) + I + G \ , \qquad (4.1)$$

where we again consider I, T and G to be fixed. We suppose that Government attempts to fix T and G so that $Y_o = Y_F$.

We have

$$d\{C(Y-T) + I + G\}/dY < 1 \ , \qquad (4.2)$$

and so $Y_o < Y_F$ implies

$$Y_F - C(Y_F - T) - I - G > 0 \ . \qquad (4.3)$$

The left-hand side of (4.3) is the <u>deflationary gap</u>. To eliminate it, the Government may increase G or decrease T or a combination of these. (Note that if G is increased and T is decreased by the same amount, then the deflationary gap is decreased.)

If, however, the values of I,T,G are such that (4.1) would give a forbidden value $Y_o > Y_F$, then there is an <u>inflationary gap</u>

$$C(Y_F - T) + I + G - Y_F > 0$$

and Government should decrease G or increase T .

Finally, suppose that, by good luck or good guidance, I,G,T have been fixed so that there is precisely full employment: $Y_o = Y_F$. Suppose also that Government feels compelled to increase G (e.g. to fight a war). Then, to avoid an inflationary gap, it must make a greater increase in T . For a small increase δ in G , the increase in T should be δ/c , where $c = C'(Y_F - T) < 1$. This is the <u>balanced budget multiplier theorem</u>.

5. <u>Prices</u>

We must now introduce money explicitly into the story. All the goods which are included in the aggregates Y,I,G etc. are exchanged for money (say £'s sterling). As the various parameters of the model change, so do their prices. We shall suppose, however, that all the prices move in step. We may therefore introduce a parameter P , the <u>level of prices</u>. We shall denote by Y, C, I, G etc. the corresponding quantities measured at constant prices (in real terms). Thus the amount of money which must be paid for the goods comprised in (say) I when the price level is P is PI.

6. Interest

Hitherto we have regarded the rate of investment I as fixed in our models. It is nowadays regarded as natural to suppose that it depends primarily on the rate of interest i, whieh we must now briefly introduce.

We suppose that there are bonds, which may be purchased by the investor and which convey the riaht to a fixed payment annually in perpetuity (as Consols do in Britain). Let £b be the price of such a bond yielding £1 per annum. The (annual) rate of interest is, by definition, $i = 1/b$.

Keynes argued that i cannot, in the real world, become arbitrarily small, because of the existence of uncertainty in our sublunary affairs. Suppose that an investor buys a bond yielding £1 per annum when the rate of interest is i_1, so the price is $1/i_1$. If at the end of the year the rate of interest is i_2, then he will have a bond worth $1/i_2$ together with £1 interest. If i_1 is already small, then the probability is that the rate of interest will increase, and so the loss $(1/i_1)-(1/i_2)$ in the value of the bond would far more than outweigh the £1 interest. This argument was christened the liquidity trap by our own Professor Dennis Robertson.

To revert to investment. It is supposed that the saver has the choice of putting his money either into bonds or into investment goods. If i is large, the bonds are a "good buy" and there will be few investments in investment goods which will hold out the prospect of an equal return. Hence we assume that investment I in such goods depends only on i and decreases as i

increases, that is

$$I = I(i) \ , \ I'(i) < 0 \ . \qquad (6.1)$$

These arguments are relevant to the "Paradox of Thrift" [cf.(2.14)]. If people become more thrifty and try to save more, the demand for bonds will increase. The price of bonds increases; that is, the interest rate falls. Now investment I increases. But $S = I$, so S increases. The effect of an increase in thrift on the gross national product Y (increase or decrease) in a model of an economy with bonds depends on the properties postulated for the model.

7. Money

We suppose that there is a certain amount of money M circulating in the economy. It has to be explained why people prefer to hold money (remain liquid), when they could invest it profitably in industry or in bonds. Bowdlerizing Keynes we distinguish two kinds of reason.

(i) Speculative and precautionary motives. We have already touched on these in the discussion of the liquidity trap. People may hold money because they think that they may be able to invest it more profitably later when there is a change in the economic climate. Alternatively, they may feel the need to be in a position to react quickly if disaster strikes unexpectedly. Denote the money held for either of these reasons by L_1. When interest rates are high, so is the temptation to buy bonds. Hence it is supposed that L_1 depends only on the rate of interest i, and decreases as i increases.

(ii) Transactions motive. People need to hold money in order to carry out their business transactions. It is reasonable to suppose that the total amount L_2 of money held for this reason depends on the amount of business to be transacted. We therefore suppose that L_2 depends only on the GNP Y and increases as Y

increases.

In the above argument it is clear that what matters is not the nominal amount of money M but the amount of money M/P in real terms (cf. §5). Hence we have

$$M/P = L_1(i) + L_2(Y) \tag{7.1}$$

or, more generally,

$$M/P = L(i,Y) \tag{7.2}$$

where

$$\partial L/\partial i < 0 \ ; \ \partial L/\partial Y > 0 \ . \tag{7.3}$$

We have already (in §3) envisaged the possibility that the demand G of government for goods etc. is not equal to the taxes T which it levies. The government must, however, pay for what it uses. If $G > T$ one possibility is that it creates ("prints") the additional money necessary. [This was not possible in the old days: "money" meant silver bullion to Adam Smith.] Alternatively, government may create and sell bonds to finance the deficit $G-T$. If $T > G$, so the government collects more than it spends, these processes are put in reverse.

8. <u>The labour market</u>

We suppose that the <u>level</u> N <u>of employment</u> (i.e. the number of people employed) and the GNP determine each other, say

$$Y = Y(N) \qquad Y' > 0 \ . \tag{8.1}$$

As already explained, we do not necessarily suppose that the labour market is in equilibrium. We denote by W the <u>wage</u> in money terms, so W/P is the <u>real wage</u>. We suppose that the demand N_D for labour depends only on the real wage, say

$$W/P = \phi(N_D) \ , \qquad \phi' < 0 \ . \tag{8.2}$$

We also suppose that the supply N_S of labour depends only on the real wage

$$W/P = \psi(N_S) \ , \qquad \psi' > 0 \ . \tag{8.3}$$

By <u>full employment</u> we mean that the labour market

clears. This occurs at a wage W and level N_F of employment such that

$$W/P = \phi(N_F) = \psi(N_F) \ . \qquad (8.4)$$

Hence the full level of employment N_F and the corresponding W/P are uniquely determined. Then (8.1) determines the full-employment GNP Y_F, which was introduced in §4, by

$$Y_F = Y(N_F) \ . \qquad (8.5)$$

9. Full employment

We now have all the ingredients for what is now called the classical model. By (3.4) we have

$$S(Y-T) - G+T = I(i) \ , \ I' < 0 \ ; \qquad (9.1)$$

and for convenience we recall (7.2),(7.3):

$$M/P = L(i,Y) \quad \partial L/\partial i < 0 \ , \ \partial L/\partial Y > 0 \ . \qquad (9.2)$$

Here M,G,T are supposed to be given. Under conditions of full employment $Y = Y_F$ is given. Then (9.1) determines the rate of interest i and, finally, (9.2) gives the level of prices P. It is fortunate that our collection of simultaneous equations can be solved in this simple manner.

We now consider the effect on the model of various changes.

(i) We note first that M occurs only in the combination M/P. Hence the quantity theory of money holds in its crudest form: if the money-supply M is doubled, then the level of prices P is doubled but nothing else is affected.

(ii) Suppose that the government increases its expenditure G but leaves taxes T and money supply M unchanged. The GNP remains unchanged at its full-employment level $Y = Y_F$. By (9.1) investment I must decrease, and so interest i increases. [We can explain this, recalling that government finances its deficit by issuing bonds: more bonds are issued, so they become cheaper.] Then M/P decreases by (9.2);

but M is fixed so P increases. This is an example of a <u>demand-pull inflation</u>.

(iii) Suppose that labour demands a higher wage, i.e. that the function ψ in (8.3) is replaced by a new and larger function of N_S but everything else is unchanged. Then the level N_F of full employment decreases and hence so does $Y = Y_F$. By (9.1) we have again an increase in i and so by (9.2) an increase in P. This is a <u>cost-pull inflation</u>.

(iv) This is a more radical change. We suppose (halcyon days!) that the labour force is unaware of the possibility of inflation, and that the supply of labour depends not on the real wage W/P but on the money wage W. (<u>Monetary illusion</u>.) Instead of (8.3) we thus have

$$W = \psi(N_S), \qquad \psi' > 0; \tag{9.3}$$

and so the level of full employment N_F and the corresponding wage now depend on the price level P, being given by

$$W/P = \phi(N_F), \qquad W = \psi(N_F). \tag{9.4}$$

Hence full-employment GNP Y_F also depends on P. The solution of the simultaneous equations (8.1),(9.1),(9.2) and (9.4) is no longer so easy. It is left to the reader to verify that an increase in G now leads to an increase in Y as well as the effects described in (ii).

10. <u>Unemployment</u>

We now suppose that unemployment is present and so the considerations of §8, which give the value of the GNP Y, no longer apply. Equations (9.1),(9.2), however, continue to hold. For fixed values of G,T,M,P each gives a relation between Y and i. We plot them on a diagram with Y horizontal and i vertical (the <u>Hicks diagram</u>, so called after the Oxford economist who introduced it).

Figure 6.

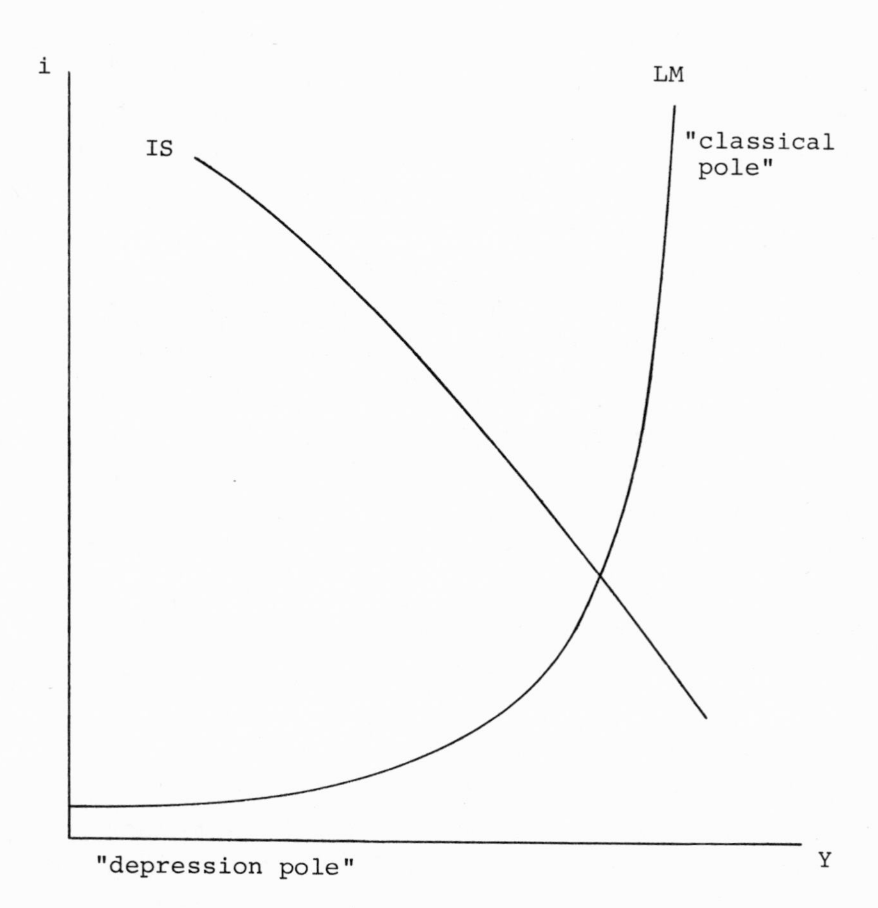

The Hicks diagram

The graph of (9.1) is traditionally called the IS-curve. Since S increases with Y but I decreases with i, the value of i on this curve decreases as Y increases. The graph of (9.2) is the LM-curve. The properties of $L(i,Y)$ imply that on it i increases as Y increases. Hence the two curves intersect (if at all) at a single point, which gives Y and i.

The liquidity trap argument of §6 implies that i tends to some $i_0 > 0$ as $Y \to 0$. There is thus an almost horizontal portion of the LM-curve when Y is small. This is called the depression pole and is said to represent the situation when there is little economic activity, as in the 1930's.

It is reasonable to assume that there is a limit to the amount of production Y which can be financed by a given amount of money M/P in real terms. Hence as Y increases the LM curve will approach a horizontal asymptote. The nearly vertical portion of the LM-curve is the classical pole. It is alleged to describe the situation described by the classical economists (i.e. before Keynes).

We are now in a position to consider the effects of changes in the parameters. We suppose that the price-level P remains fixed and distinguish:

Monetary changes, i.e. changes in M. These affect only the position of the LM-curve. An increase in M moves it to the right, roughly speaking.

Fiscal changes, i.e. changes in T and/or G. These affect only the IS-curve. An increase in G (or a decrease in T) moves it approximately to the right.

We also distinguish cases according to the position of the LM-curve of its intersection with the IS-curve.

Classical pole. Here an increase of M increases Y and decreases i slightly. On the other hand, fiscal changes have hardly any effect on Y but an increase in G (or a decrease in T) increases i.

Depression pole. A change in M has little effect on either Y or i. On the other hand, an increase in G (or a decrease in T) increases Y substantially but increases i only minimally.

11. The long term

Investment has played a key role in the earlier part of this chapter but it has been a singularly unmotivated activity. Its purpose is, of course, to maintain or increase the community's capital stock K. In the long run, changes in K will affect the functional relationships between the other parameters in the economy, e.g. by reducing the labour required for a given output. One might ask, for example, how one should best allot production between consumption and investment so as to maximize the welfare of the community over all future time: if too much is consumed then future generations will have an inadequate stock of capital but it is equally possible to invest too much and so provide inadequately for present enjoyment. Pioneering work on this question was done by the young Cambridge mathematician Frank Ramsey. To give any reasonable introduction to these questions would go far beyond the modest aims of this book (not to mention the competence of its author), and we shall confine ourselves to a couple of 5-finger exercises which give something of the flavour.

For our first model we suppose that the capital stock is indestructible and never gets out of date. If $K(t)$ is the capital stock and $I(t)$ the investment at time t, then

$$K(t+1) = K(t) + I(t) . \qquad (11.1)$$

We suppose that output depends solely on the capital stock, say

$$K(t) = vY(t) \qquad (11.2)$$

for some constant v. Suppose, also, that saving (and

so investment) is proportional to income, say

$$I(t) = sY(t) \ . \qquad (11.3)$$

On eliminating $Y(t)$, $I(t)$ from (11.1)-(11.3) we have

$$K(t+1)/K(t) = (1+g_o) \ , \qquad (11.4)$$

where

$$g_o = s/v \qquad (11.4 \text{ bis})$$

is Harrod's <u>warranted rate of growth</u>. The whole economy thus expands at the rate $(1+g_o)$. Now suppose that production also depends on labour N, but only in the sense that units of capital require a constant amount of manning. This leads to the production function

$$Y = \min(N/u, K/v) \qquad (11.5)$$

for constants u, v. Suppose also that N has a natural rate of growth, say

$$N(t+1) = (1+g)N(t) \ . \qquad (11.6)$$

If $g > g_o$, the GNP $Y(t)$ will increase at the warranted rate but there will be unemployment. If, however, $g < g_o$, then the GNP can increase only at the slower rate $(1+g)$, and capital is accumulated which cannot be used.

In our second model it is convenient to suppose that the capital stock is consumed every year and so the new capital stock must be taken from each year's production. (It is seed corn rather than blast furnaces.) Output depends only on the capital K and the labour N employed. There are constant returns to scale, so

$$Y = Nf(k) \ , \qquad k = K/N \qquad (11.7)$$

where k is the <u>capital intensity</u>. The marginal product of labour

$$\partial Y/\partial N = f(k) - kf'(k) \qquad (11.8)$$

is the wage. The marginal product of capital

$$\partial Y/\partial K = f'(k) \qquad (11.9)$$

gives the rate of interest $f'(k)-1$, since a saving of one unit at time t gives an output of $f'(k)$ at time $t+1$.

We now suppose that the population grows at a constant rate:

$$N(t+1) = (1+g)N(t) \tag{11.10}$$

and that the economy expands perfectly steadily so that the per capita capital k and the per capita consumption c remain constant. The output Y is divided between capital and consumption:

$$\begin{aligned} Y(t) &= K(t+1)+C(t+1) \\ &= (k+c)N(t+1) \\ &= (1+g)(k+c)N(t) \ . \end{aligned} \tag{11.11}$$

Hence

$$(1+g)(k+c) = f(k) \ . \tag{11.12}$$

The community, if it is wise, will allocate its income between consumption and saving so as to maximize c ; that is $k = k^*$, where

$$f'(k^*) = 1+g \ . \tag{11.13}$$

This is the golden rule: that the rate of interest should be equal to the rate of growth.

Since Y is homogeneous, we have

$$Y = N\partial Y/\partial N + K\partial Y/\partial K \ . \tag{11.14}$$

It is readily verified that at the golden rule capital intensity $k = k^*$ one has

$$(N\partial Y/\partial N)(t) = C(t+1) \tag{11.15}$$

$$(K\partial Y/\partial K)(t) = K(t+1) \ . \tag{11.16}$$

Here (11.15) says that consumption is equal to the total amount of wages and, similarly, (11.16) states that investment equals the profits of capital.

DEFINITION OF SYMBOLS

C : consumption (§2) at constant prices (§5)

c : marginal propensity to consume (§2). But used for per capita consumption in §11

G : Government demand (§2) at constant prices (§ 5)

g : the rate of expansion in §11 is $1+g$

I : investment (§2) at constant prices (§5)

i : rate of interest (§6)

K : capital stock (§11)

k : capital intensity (per capita amount of capital)(§11)

L : the liquidity function (§7)

M : money supply (§7)

N : level of employment (§8)

N_F : level of full employment (§8)

P : level of prices (§5)

Q : total output. Identified with Y in steady state (§2)

S : saving (§2) at constant prices (§5)

s : marginal propensity to save (§2)

T : taxes (§3)

t : index denoting time

W : wage (in money terms) (§8)

Y : total income of factors of production. Referred to as GNP (= gross national product) (§2). It is valued at constant prices (§5)

Y_F: full-employment GNP (§§4,8)

Z : aggregate demand. Identified with Y in steady state (§2)

Further reading

R.G.D. Allen. *Macro-economic theory* (Cambridge U.P., 1967).

E. Burmeister. *Capital theory and dynamics* (Cambridge U.P., 1980).

R.L. Crouch. *Macroeconomics* (Harcourt, Brace, Jovanovich, 1972).

G. Hadley and M.C. Kemp. *Variational methods in economics* (North Holland, 1971).

P.A. Samuelson. *Economics* (McGraw Hill, Kogakusha, n-th edition, $n \to \infty$).

S.J. Turnovsky. *Macroeconomic theory and stabilization policy* (Cambridge U.P., 1977).

Chapter 6 Exercises

1. Consider the effect of changes of thrift (as defined in §2) on the model described in §10. Similarly for changes in the "propensity to invest" $I(i)$.

2. A certain country has an entirely agricultural economy. All prices, wages etc. are expressed in terms of agricultural produce. The total number of workers is W . There are a number of village sites V_i . If there are t workers in V_i , then they produce $f_i(t)$ of agricultural produce, where

$$f_i(0) = 0 \text{ , } f'(t) > 0 \text{ , } f''(t) < 0 \ (t \geq 0) \text{ .}$$

We consider several different types of economy.

(i) (Primitive.) There are x_i workers at V_i and they each receive $f_i(x_i)/x_i$. They migrate until this is equalized. Show that there is a constant X such that

$$x_i = 0 \quad \text{if} \quad f_i'(0) \leq X$$
$$f_i(x_i) = Xx_i \quad \text{otherwise.}$$

(ii) (Socialist.) Here y_i workers are assigned to the village V_i so as to maximize total output, which is then equally divided among all the workers. Show that there is a constant Y such that

$$y_i = 0 \quad \text{if} \quad f_i'(0) \leq Y$$
$$f'(y_i) = Y \quad \text{otherwise.}$$

Show that the wage is greater than under primitive conditions and that in general some villages are populated which are not populated in the primitive state.

(iii) (Capitalist.) Each village has a landlord (who is not a worker). Landlords pay workers a wage Z (in agricultural produce), the same for all villages. The landlord of village V_i selects the number z_i of workers to maximize his profit

$$R_i = f_i(z_i) - Zz_i = \max_t\{f_i(t) - Zt\} .$$

He uses the profits to employ workers to make luxuries, paying them the wage Z , so that the total number ℓ of workers in the luxury trades is given by

$$\ell Z = \Sigma R_i .$$

The wage Z is determined by the condition

$$\ell + \Sigma z_i = \text{total work force } (= W) .$$

Show that

(α) $X \geq Z$ (wage under capitalism is less than primitive wage).

(β) In general more villages are populated under capitalism than under primitive society, but fewer than under socialism.

(γ) Total agricultural production $\Sigma f_i(z_i)$ under capitalism is less than primitive production $\Sigma f_i(x_i)$.

(δ) Total production of agriculture plus luxuries (valued at labour content) under capitalism is greater than total primitive production. [J.S. Cohen and M.L. Weizman in: Mathematical models in economics (J. and M.W. Łos, editors) 1974.]

3. (Inventory cycle.) (i) A manufacturer makes a good G . The process takes a year, and the only input is the good G itself. The production of a unit of G (net) requires the input of λ units of G . Let $b(t)$ be the stock (American: inventory) of G held at the beginning of year t and let $a(t)$ be amount of G manufactured in that year, so $0 \leq a(t) \leq \lambda^{-1}b(t)$. At the end of each year a quantity e (independent of t) is consumed. Show that

$$b(t+1) = b(t) + a(t) - e .$$

(ii) Each year the manufacturer endeavours to bring his stock up to an amount which would have sufficed to produce the amount manufactured in the previous year, together with an allowance for contingencies. He thus aims at a stock $\ell a(t-1) + m$, where $\ell > \lambda$ and $m > e$ are fixed. Show that he manufactures $a(t)$, where

$$a(t) = \begin{cases} 0 & \text{if } a^*(t) < 0 , \\ \lambda^{-1}b(t) & \text{if } a^*(t) > \lambda^{-1}b(t) , \\ a^*(t) & \text{otherwise} , \end{cases}$$

and

$$a^*(t) = \ell a(t-1) + m - b(t) .$$

(iii) Show that there is a single fixed point $(\bar{a},\bar{b})$, in the sense that $a(t) = \bar{a}$, $b(t) = \bar{b}$ implies $a(t+1) = \bar{a}$, $b(t+1) = \bar{b}$.

(iv) If $\ell > 1$, show that the fixed point $(\bar{a},\bar{b})$ is unstable. For certain values of the parameters λ,ℓ,m, show that $(a(t),b(t))$ remains bounded as $t \to \infty$ and is infinitely often on each of the "slump line" $a(t) = 0$ and the "boom line" $a(t) = \lambda^{-1}b(t)$. For other values of the parameters, show that $a(t) \to \infty$, $b(t) \to \infty$ along the boom line.

[Hint. If $(a(t),b(t))$ and $(a(t+1),b(t+1))$ are not on the slump or boom lines, show that $\alpha(t) = a(t) - \bar{a}$, $\beta(t) - \bar{b}$ satisfy

$$\alpha(t+1) = (\ell-1)\alpha(t) - \beta(t)$$

$$\beta(t+1) = \alpha(t) + \beta(t) .$$

Note. The scenario gains in plausibility if there are several goods produced in a closed Leontieff model (cf. Chapter 5, §2). For mathematical tractability one is then led to make the (implausible) hypothesis that all occurring bundles of commodities are multiples of the Sraffa bundle. This comes back to the model above. See J.T. Schwartz, Lectures on the mathematical method in analytical economics, and, for an extension, his Theory of money.]

4. Verify (11.15) and (11.16).

[For generalization cf. E.S. Phelps, Second essay on the golden rule of accumulation. Amer. Econ.Rev. 55 (1965), 793-814.]

5. Let $Y^* = PY$ be the GNP in money terms and similarly for S^*, I^* etc. Consider a model of the world in which S^* depends on $Y^* - T^*$, I^* depends on i and the demand for money is a function $L^*(i, Y^*)$. On replacing (9.1),(9.2) by

$$S^*(Y^* - T^*) - G^* + T^* = I^*(i) ,$$

$$M = L^*(i, Y^*) ,$$

show that the nominal GNP Y^* is determined for fixed G^*, T^* by the money supply M.

[Reference. F. Modigliani. Econometrica, 12 (1944), 45-92.]

APPENDIX A

Convex Sets

1. Fundamentals

Definition. A set $C \subset R^n$ is convex if

$$\lambda \underline{c}_o + (1-\lambda)\underline{c}_1 \in C \qquad (0 \le \lambda \le 1)$$

whenever $\underline{c}_o, \underline{c}_1 \in C$.

Corollary. Let $K > 1$ and let $\underline{c}_1, \ldots, \underline{c}_K \in C$. Then

$$\sum_{k=1}^{K} \lambda_k \underline{c}_k \in C \qquad (\lambda_k \ge 0,\ \Sigma\lambda_k = 1) .$$

Lemma 1. Suppose that C is convex. Then the interior C^o and the closure $\bar{C}$ are convex.

Proof. Clear.

Lemma 2. Let C be convex. Then the two following statements are equivalent.

(i) the interior C^o is not empty;

(ii) for any $\underline{c}_o \in C$ there are $\underline{c}_1, \ldots, \underline{c}_n \in C$ such that $\underline{c}_j - \underline{c}_o$ $(1 \le j \le n)$ are linearly independent.

Proof. (i) → (ii) . Trivial without hypothesis of convexity.

(ii) → (i) . By the Corollary to the definition $(K = n+1)$ the set C contains the simplex with vertices $\underline{c}_o, \underline{c}_1, \ldots, \underline{c}_n$.

Lemma 3. Let C be convex. If C^o is not empty, then its closure is $\bar{C}$.

Proof. Clear.

Theorem 1. Let C be convex and open. Let $\underline{b} \notin C$. Then there is a hyperplane H through $\underline{b}$ which does not meet C .

Note. Hence C lies entirely on one side of H .

Proof. We may suppose without loss of generality that

$$\underline{b} = \underline{0}$$

is the origin.

n = 2. We say that $r \in A^+$ if there is a $t > 0$ such

that $(t,rt) \in C$. Clearly A^+ is non-empty and open. Similarly $r \in A^-$ if there is an $t < 0$ such that $(t,rt) \in C$; and A^- is non-empty and open. If $r \in A^+ \cap A^-$, we should have $\underline{0} \in C$ by convexity: so A^+, A^- are disjoint. Since the real line is connected, there is thus an $r \notin A^+, \notin A^-$. Then the line (t,rt) $(-\infty < t < \infty)$ does not meet C.

<u>$n > 2$</u>. By considering any plane through $\underline{0}$ and its intersection with C, there is certainly a line L through $\underline{0}$ which does not meet C. Consider the projection

$$\pi : R^n \to R^n/L \cong R^{n-1} .$$

The projection πC is clearly open and convex, and it does not meet the origin of R^n/L by the construction of L. By induction, there is a hyperplane P of R^n/L which does not meet πC. Then $H = \pi^{-1}P$ is a hyperplane of R^n not meeting C. This concludes the proof.

Let now C be convex with non-empty interior and let $\underline{f}$ be a point of the frontier of C (i.e. $\underline{f} \in \bar{C}; \underline{f} \notin C^o$). By Theorem 1 applied to C^o there is a hyperplane H through $\underline{f}$ which does not meet C^o. It is called a <u>tac-hyperplane</u> to C at $\underline{f}$ and is not necessarily unique. By Lemma 3, C lies entirely in one of the two closed half-spaces defined by H. When the frontier of C has a tangent hyperplane at $\underline{f}$, then it is clearly the unique tac-hyperplane at $\underline{f}$.

<u>Lemma 4</u>. Let C_λ $(\lambda \in \Lambda)$ be a family of convex sets in R^n. Then

$$\bigcap_\lambda C_\lambda$$

is also convex.

<u>Proof</u>. Follows from definition of convexity.

Let S be any set in R^n, and let

$$C = \cap D \qquad (D \text{ convex}, D \supset S) .$$

Then C is convex (by Lemma 4) and $C \supset S$. It is called the <u>convex cover</u> (American: convex hull) of S.

In a strict sense it is the smallest convex set which contains S .

Lemma 5. The convex cover of a set S is the set of points $\underline{c}$ which can be represented as

$$c = \sum_{k=1}^{K} \lambda_k \underline{s}_k \qquad (*)$$

for some

$$K > 0 \ , \ \underline{s}_k \in S \ , \ \lambda_k \geq 0 \ , \ \Sigma\lambda_k = 1 \ .$$

Proof. Any convex set which contains S must contain $\underline{c}$. Further, the set C^* of points of the shape (*) is readily verified to be convex. Hence C^* is the convex cover.

2. Separation theorems

In this section we show that a pair of disjoint convex sets can (in some sense) be separated by hyperplanes. There are a number of variants, and in the text we shall mainly require only one application (Lemma 8).

The following Lemma will be superseded by Theorem 2.

Lemma 6. Let $C \subset R^n$ be convex and suppose that $\underline{0} \notin C$. Then there is a $\underline{v} \neq \underline{0}$ such that

$$\underline{vc} \geq 0 \qquad (\text{all } \underline{c} \in C) \ . \qquad (£)$$

Proof. (i) Suppose, first, that the interior C^o of C is not empty. By Theorem 1 there is a hyperplane $\underline{vx} = 0$ through $\underline{0}$ which does not meet C^o; and so on taking $-\underline{v}$ for $\underline{v}$ if need be we have

$$\underline{vx} > 0 \qquad (\text{all } \underline{x} \in C^o) \ .$$

Then (£) follows from Lemma 3.

(ii) Otherwise, C^o is empty and so, by Lemma 2, there is a $\underline{w} \neq \underline{0}$ and q such that C lies entirely in the hyperplane

$$H : \underline{wx} + q = 0 \ .$$

If $q \neq 0$, we can take $\underline{v} = \pm \underline{w}$. If, however, $q = 0$, then the result follows by induction on n , on considering the (n-1)-dimensional linear space H .

If $S_1, S_2 \subset R^n$, then we denote by $S_1 - S_2$ the set of

$\underline{x}$ which can be put in the form $\underline{x} = \underline{s}_1 - \underline{s}_2$ $(\underline{s}_j \in S_j)$.

Lemma 7. Let $C_1, C_2 \subset R^n$ be convex. Then

$$C = C_1 - C_2$$

is convex.

Proof. Follows at once from the definition of convexity.

Theorem 2. Let C_1, C_2 be convex and disjoint. Then there is a $\underline{v} \neq \underline{0}$ and a q such that

$$\underline{vc}_1 \geq q \qquad (\text{all } \underline{c}_1 \in C_1)$$

$$\underline{vc}_2 \leq q \qquad (\text{all } \underline{c}_2 \in C_2) .$$

Proof. $C = C_1 - C_2$ is convex by Lemma 7, and so by Lemma 6 there is a $\underline{v} \neq \underline{0}$ such that

$$\underline{v}(\underline{c}_1 - \underline{c}_2) \geq 0 \qquad (\text{all } \underline{c}_j \in C_j) .$$

Hence

$$\inf \underline{vc}_1 \geq \sup \underline{vc}_2 ,$$

and the existence of q follows.

Note. Lemma 6 is the special case in which C_2 is a single point.

A set $K \subset R^n$ is called a cone if $\underline{k} \in K$ implies $\lambda\underline{k} \in K$ for all $\lambda > 0$. We denote by Ω the open positive orthant in R^n :

$$\Omega = \{\underline{x} : \underline{x} >> 0\} .$$

Lemma 8. Let K be a convex cone in R^n . Then precisely one of the two following statements holds:

(i) $K \cap \Omega$ is not empty.

(ii) There is a $\underline{p} > \underline{0}$ such that

$$\underline{pk} \leq 0 \qquad (\text{all } \underline{k} \in K) .$$

Proof. It is clear that (i) and (ii) cannot hold simultaneously. Suppose that (i) is false. Then by Theorem 2 there is a $\underline{v} \neq 0$ and a q such that

$$\underline{vx} \geq q \qquad (\text{all } \underline{x} \in \Omega) \qquad (@)$$

$$\underline{vx} \leq q \qquad (\text{all } \underline{x} \in K) .$$

Now (@) implies readily that $\underline{v} > \underline{0}$ and then that $q \leq 0$. Hence $\underline{p} = \underline{v}$ satisfies (ii) .

Corollary. Let the matrix D have n rows and m columns. Then precisely one of the two following statements holds:

(i) $D\underline{t} \gg \underline{O}$ for some $\underline{t} > \underline{O}$ in R^m.

(ii) $\underline{p}D \le \underline{O}$ for some $\underline{p} > \underline{O}$ in R^n.

Proof. We can take

$$K = \{\underline{x}: \underline{x} = D\underline{t} \text{ for some } \underline{t} > \underline{O}\}.$$

Although we do not need the results, there is some interest in pursuing the ideas further.

Lemma 9. Let C be both convex and closed. Suppose that $\underline{O} \notin C$. Then there is a $\underline{v} \neq \underline{O}$ and a $\delta > 0$ such that

$$\underline{vc} \ge \delta \qquad \text{(all } \underline{c} \in C).$$

Proof. Since C is closed, there is an $\eta > 0$ such that the closed spherical ball B of radius η and centre $\underline{O}$ does not meet C. By Theorem 2 with $C_1 = C$, $C_2 = B$ there is a $\underline{v}$ and a q such that

$$\underline{vc} \ge q \qquad \text{(all } \underline{c} \in C)$$

where

$$q \ge \sup \underline{vs} \qquad (\underline{s} \in S)$$
$$> 0.$$

Hence $\delta = q$ will do.

Note. If $\underline{O} \notin C$, where C is convex and is either open or closed, there is a $\underline{v} \neq \underline{O}$ such that $\underline{vc} > 0$ for all $\underline{c} \in C$ by Theorem 1 (C open) or Lemma 10 (C closed). If C is neither open or closed, no such $\underline{v}$ need exist. Consider for example $n = 2$ where C consists of the points with $0 < x_1 < 1$, $|x_2| < 1$ together with $(x_1,x_2) = (0,1)$.

Theorem 3. Let C_1, C_2 be a disjoint pair of closed convex sets. Suppose that at least one of C_1, C_2 is bounded. Then there is a $\underline{v} \neq \underline{O}$ such that

$$\inf \underline{vc}_1 > \sup \underline{vc}_2 \qquad (\underline{c}_j \in C_j).$$

Proof. The conditions imply that $C = C_1 - C_2$ is closed. The proof now follows that of Theorem 2, but uses Lemma 10 instead of Lemma 6.

Note. The conclusion may cease to hold if C_1, C_2 are both allowed to be unbounded. Consider $n = 2$ and

$$C_1 : x_1 > 0 \ , \ x_2 > 0 \qquad\qquad x_1x_2 \geq 1$$
$$C_2 : x_1 \geq 0 \ , \ x_2 \leq 0 \ .$$

There is not even a $\underline{v} \neq \underline{0}$ such that $\underline{vc}_1 > 0$ and $\underline{vc}_2 < 0$ for all $\underline{c}_j \in C_j$.

3. Differential properties

We require this section only for $n = 1$. However, the general case illuminates some of the other results which are obtained in other ways, and can be used to provide alternative proofs of them.

Definition. Let $C \subset R^n$ be convex and have non-empty interior C^o . A function $f : C \to R$ is said to be convex if the (n+1)-dimensional set

$$(\underline{x},y) \qquad \underline{x} \in C \qquad y \geq f(\underline{x})$$

is convex. This is obviously equivalent to the condition

$$f(\lambda\underline{x}_o + (1-\lambda)\underline{x}_1) \leq \lambda f(\underline{x}_o) + (1-\lambda)f(\underline{x}_1)$$

for

$$\underline{x}_o \ , \ \underline{x}_1 \in C \ , \qquad 0 \leq \lambda \leq 1 \ .$$

Theorem 4. Suppose that f has continuous second derivatives

$$f_{ij}(\underline{x}) = \partial^2 f/\partial x_i \partial x_j \ .$$

Then the two following statements are equivalent

(i) f is convex ;

(ii) the quadratic form

$$Q(\underline{x})(X_1,\ldots,X_n) = \Sigma f_{ij}(\underline{x})X_1X_j$$

is positive definite or semi-definite for all $\underline{a} \in C$.

Note. We shall be concerned only with the case $n = 1$, when $\underline{x} = x$ is scalar and (ii) becomes simply

$$f''(x) \geq 0 \ .$$

Proof. $n = 1$, (i) $\to$ (ii) . Let $x_o \in C^o$ and $y_o = f(x_o)$. The tangent

$$y - y_o = f'(x_o)(x-x_o)$$

is a tac line, so

$$f(x) - f(x_o) \geq f'(x_o)(x-x_o)$$

for all $x \in C$. On letting $x \to x_o$, we get $f''(x_o) \geq 0$. By continuity, $f''(x) \geq 0$ for all $x \in C$.

<u>$n = 1$, (ii) $\to$ (i)</u>. We have to show that

$$f(\lambda x_o + (1-\lambda)x_1) \leq \lambda f(x_o) + (1-\lambda)f(x_1) \qquad (\S)$$

whenever

$$x_o, x_1 \in C , \qquad 0 < \lambda < 1$$

and, without loss of generality,

$$x_o < x_1 .$$

The difference between the two sides of (§) is

$$\lambda\{f(\lambda x_o + (1-\lambda)x_1) - f(x_o)\}$$
$$+ (1-\lambda)\{f(\lambda x_o + (1-\lambda)x_1) - f(x_1)\} .$$

By the Mean Value Theorem applied to each bracket, this is

$$\lambda(1-\lambda)(x_1-x_o)\{f'(\xi_1) - f'(\xi_2)\}$$

where

$$x_o < \xi_1 < \lambda x_o + (1-\lambda)x_1 < \xi_2 < x_1 .$$

A second application of the Mean Value Theorem gives

$$f'(\xi_1) - f'(\xi_2)$$
$$= (\xi_1 - \xi_2)f''(\xi_3) \qquad \xi_1 < \xi_3 < \xi_2$$
$$\leq 0 ,$$

as required .

<u>$n > 1$</u>. Follows from $n = 1$ by considering the restriction of f to $C \cap L$, where L is any line in R^n.

Appendix A Exercises

1. For $S, T \subset R^n$ show that

$$\mathrm{con}(S+T) = \mathrm{con}(S) + \mathrm{con}(T),$$

where $\mathrm{con}(S)$ is the convex cover of S and $S+T$ is the set of $\underline{s}+\underline{t}$ $(\underline{s} \in S, \underline{t} \in T)$.

2. (Shapley-Folkman). Let $S_t \subset R^n$ $(1 \le t \le T)$, and let

$$\underline{c} \in \mathrm{con}(\sum_t S_t).$$

Show that

$$\underline{c} = \sum \underline{b}_t,$$

where $\underline{b}_t \in \mathrm{con}(S_t)$ for all t and $\underline{b}_t \in S_t$ except for at most n values of t.

[Hint.

$$\underline{c} = \sum_t \sum_{1 \le j \le J(t)} \lambda_{tj} \underline{s}_{tj}$$

for some $J(t)$, some $\underline{s}_{tj} \in S_t$ and some $\lambda_{tj} > 0$ with $\sum_j \lambda_{tj} = 1$. Choose a representation for which $\sum_t J(t)$ is minimal. If $\sum J(t) > T+n$, then the vectors

$$\underline{s}_{tj} - \underline{s}_{t1} \quad (1 \le t \le T, \ 2 \le j \le J(t))$$

of R^n are linearly dependent. Use this to eliminate one of the $\underline{s}_{tj}$, contrary to the minimality hypothesis. cf. Cassels, Math.Proc.Camb.Philos. Soc. 78 (1975), 433-436.]

APPENDIX B

The Brouwer fixed point theorem

Theorem (Brouwer). Let f be a continuous map of a closed n-dimensional simplex S into itself. Then f has a fixed point: that is, there is an $\underline{s} \in S$ such that $f(\underline{s}) = \underline{s}$.

This is a topological theorem, and so holds for any topological space homeomorphic to a simplex, e.g. a closed ball. We have, however, enunciated it in the form in which it is used in the text, and in which it is proved below. We take S to be the convex cover of the unit points in R^{n+1} : i.e. the set of

$$\underline{x} = (x_o, \ldots, x_n) \in R^{n+1} \tag{1}$$

such that

$$S : x_j \geq 0 \qquad (0 \leq j \leq n) \; ; \; \Sigma x_j = 1 \; . \tag{2}$$

Let us consider a situation whose relevance will appear only later. Denote by V_λ $(\lambda \in \Lambda)$ a simplicial decomposition of the (n+1)-dimensional set

$$U : x_j \geq 0 \qquad (0 \leq j \leq n) \; ; \; \Sigma x_j \geq 1 \; . \tag{3}$$

[That is, the V_λ are (n+1)-dimensional simplices, $U = \cup_\lambda V_\lambda$, and two of the V_λ intersect, if at all, in an r-dimensional face of each with $r \leq n$. The diagram illustrates $n = 1$.] The n-dimensional faces of the V_λ will be called facets. We make the additional hypothesis that S is a facet. Every vertex $\underline{v}$ of the V_λ has a label

$$\phi(\underline{v}) \in \{0,1,\ldots,n\} \; . \tag{4}$$

The label is subject to the condition

$$\phi(v) = i \Rightarrow v_i > 0 \; , \tag{5}$$

where v_i is the ith coordinate: but is otherwise arbitrary.

A facet F of the decomposition has $n+1$ vertices.

Figure 7.

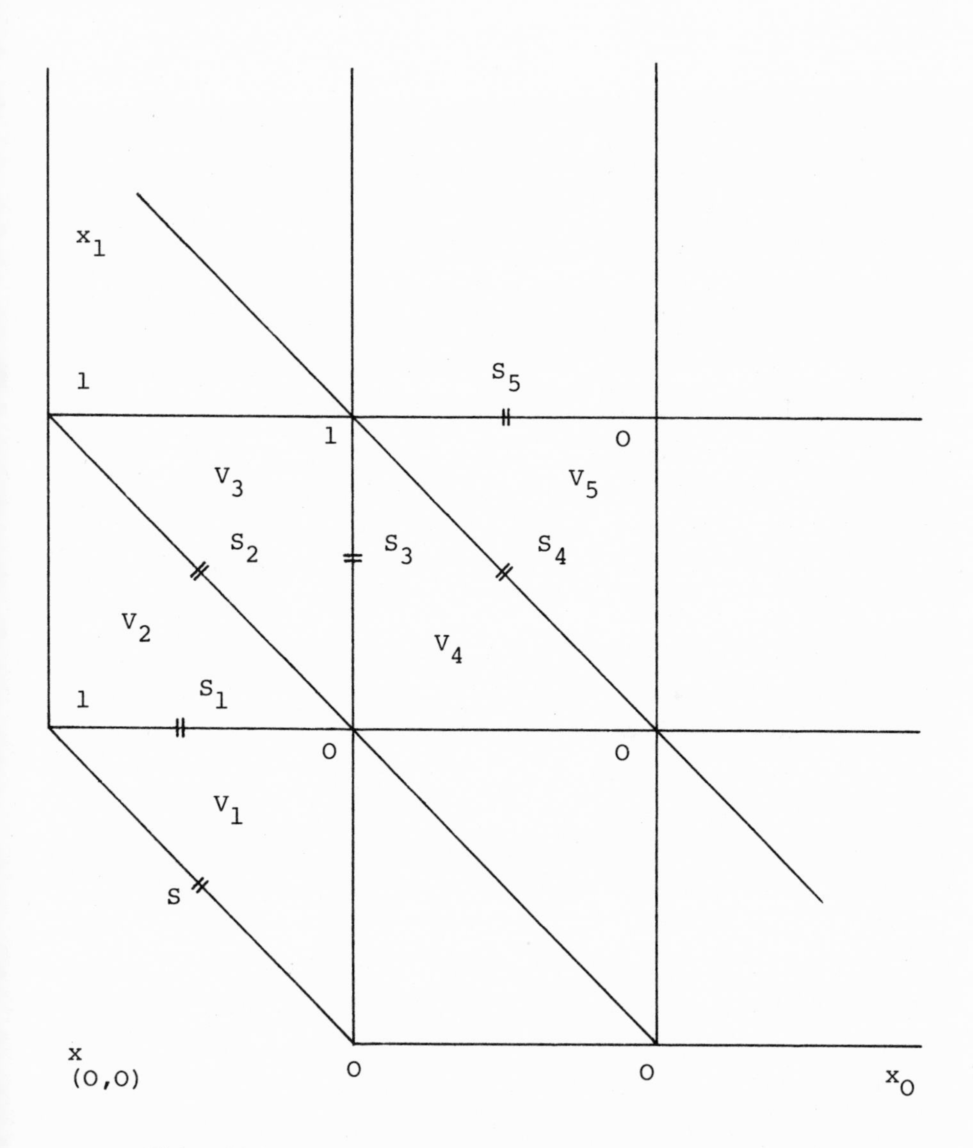

This illustrates the case $n = 1$ of Lemma 1. The completely labelled facets produced by the proof algorithm are marked.

It is said to be completely labelled if the labels of the vertices are the complete set $\{0,1,\ldots,n\}$.

Lemma 1. There are infinitely many completely labelled facets.

Proof. The condition (5) uniquely determines $\phi(\underline{v})$ for the vertices of S , and S is completely labelled. Any facet $F \neq S$ which is on the boundary of U lies on one of the hyperplanes $x_i = 0$, and so by (5) cannot be completely labelled.

If ϕ maps the $n+2$ vertices of a simplex V_λ on to the complete set $\{0,1,\ldots,n\}$, then V_λ has precisely two completely labelled facets. Otherwise it has no completely labelled facets.

The completely labelled facet S belongs to precisely one simplex V_1 (say) . Hence V_1 has precisely one other completely labelled facet S_1 . Then S_1 is not on the boundary of U , and so is a facet of precisely one other simplex V_2 . Now V_2 has precisely one completely labelled facet $S_2 \neq S_1$. In this way we obtain a uniquely determined sequence S, $S_1, S_2, \ldots$ of completely labelled facets. Further, S_k determines the pair S_{k+1}, S_{k-1} uniquely; and so the S_k $(k = 1,2,\ldots)$ are distinct. This proves the Lemma.

We shall now make two further assumptions:

(α) the vertices of the V_λ have integral coordinates

(β) every V_λ is contained in a cube of unit side.

Much weaker conditions than (α),(β) would have sufficed for our purposes. A proof that such a simplicial decomposition exists is reserved until the end of this Appendix. Denote by $\pi(\underline{u})$ the projection of $\underline{u} \in U$ on to S with centre the origin, i.e.

$$\pi(\underline{u}) = \mu\underline{u} \in S , \tag{6}$$

where $\mu = (\Sigma u_j)^{-1}$.

Corollary. Suppose that (α),(β) hold and let $\delta > 0$

be arbitrarily small. Then the diameter of $\pi(T)$ is greater than δ for only finitely many facets T. [The diameter of a set is the supremum of the distance between any two of its points.]

Proof. Clear.

We now revert to the proof of the Theorem. We label the vertices $\underline{v}$ of the simplicial decomposition V_λ as follows. Put

$$\underline{s} = \pi(\underline{v}) \in S . \tag{7}$$

Then $\phi(\underline{v})$ is any index i for which

$$f_i(\underline{s}) < s_i . \tag{8}$$

Since

$$\sum_j f_j(\underline{s}) = \sum s_j = 1 , \tag{9}$$

there is certainly such an i unless $f(\underline{s}) = \underline{s}$: when $\underline{s}$ would be a fixed point and we should be done. Clearly this labelling satisfies the condition (5).

Let T be a completely labelled facet, say with vertices $\underline{v}^{(j)}$ such that $\phi(\underline{v}^{(j)}) = j$ $(0 \le j \le n)$. Put

$$\underline{s}^{(j)} = \pi(\underline{v}^{(j)}) \in S . \tag{10}$$

Then

$$f_j(\underline{s}^{(j)}) < s_j^{(j)} \tag{11}$$

by the definition of ϕ. Let $\underline{\xi} \in S$ be any point in the convex cover of the $\underline{s}^{(j)}$. Then by the continuity of f,

$$f_j(\underline{\xi}) < \xi_j + \varepsilon \qquad (0 \le j \le n) \tag{12}$$

for any given ε, provided $\pi(T)$ is small enough: and such T exist by the Corollary. But

$$\sum f_j(\underline{\xi}) = \sum \xi_j = 1 , \text{ and so}$$

$$|f_j(\underline{\xi}) - \xi_j| \le n\varepsilon \qquad (1 \le j \le n) ; \tag{13}$$

that is, $\underline{\xi}$ is approximately fixed by f.

As $\varepsilon \to 0$, we obtain points $\underline{\xi}(\varepsilon)$. They must have a limit point $\underline{\xi}^*$. Clearly $f(\underline{\xi}^*) = \underline{\xi}^*$, and we have the required fixed point.

It remains to show that there is a simplicial decomposition V_λ of U with all the stated properties.

Lemma 2. Let

$$W : 0 \le x_j \le 1 \qquad (1 \le j \le n) \tag{14}$$

be the unit cube in R^n and let Z be the simplex

$$Z : x_j \ge 0 , \Sigma x_j \le 1 . \tag{15}$$

Then there is a simplicial decomposition of W such that (a) the vertices of the decomposition are vertices of W and (b) Z is one of the simplices.

Proof. Let W_n be the face $x_n = 0$ of W. By induction on n there is a decomposition of W_n of the required kind into (n-1)-dimensional simplices. For $1 \le j \le n-1$ let W_j be the face $x_j = 1$ of W. By induction, there is a simplicial decomposition of W_j whose vertices are the vertices of W_j. Every (n-1)-dimensional simplex in the decomposition of $W_1, W_2, \ldots, W_n$ gives an n-dimensional simplex on taking the convex cover of the union of itself and the point $(0,\ldots,0,1)$. This clearly gives the required simplicial decomposition of W.

Corollary. There is a simplicial decomposition of R^n whose vertices are the points with integral coordinates, one of the simplexes being Z.

Proof. Extend the decomposition of W to the whole of R^n by repeatedly taking the mirror image in the faces of W. Alternatively, one can arrange that the decomposition of Lemma 2 induces the same decomposition on opposite faces of W.

Finally, we get the required decomposition of U by taking $n+1$ for n in Lemma 2, Corollary and restricting to U.

In conclusion, we note that the above proof gives an efficient algorithm for finding "almost fixed" points of f. We have used the account of G. Debreu, which employs ideas of Scarf and Eaves.

References

G. Debreu. Four aspects of the mathematical theory of economic equilibrium. Proc.Intern.Con. Math. Vancouver, 1974, 65-77.

B.C. Eaves. Properly labelled simplexes. In Studies in optimization (Eds. G.B. Dantzig and B.C. Eaves). MAA studies in Mathematics, 10, 1974.

H. Scarf (with the collaboration of T. Hansen). The computation of economic equilibria. Yale Univ. Press, 1973.

Addendum. Kakutani's theorem

This is a generalization of Brouwer's theorem. Let C be a bounded closed convex set in R^n. For each $\underline{c} \in C$ let $F(\underline{c}) \subset C$ be a non-empty closed convex set (which may be of lower dimension, or indeed only a single point). The correspondence $\underline{c} \to F(c)$ is upper semi-continuous: that is the set of points $(\underline{c},\underline{f}), \underline{f} \in F(\underline{c})$ is closed in $C \times C$. Then Kakutani's theorem asserts that there exists a $\underline{c}^* \in C$ with $\underline{c}^* \in F(\underline{c}^*)$.

It is enough to consider the case of a simplex. For choose any simplex $S \supset C$. We may extend the definition of F to S by putting $F(\underline{s}) = F(\underline{c}(\underline{s}))$, where $\underline{c}(\underline{s})$ is the point of C nearest to $\underline{s}$ (so $\underline{c}(\underline{s}) = \underline{s}$ if $\underline{s} \in C$). The extended correspondence has the same fixed points as the original one.

We show here that the proof given above of Brouwer's theorem can be modified so as to give a proof of Kakutani's theorem.

Lemma 3. Let $\underline{z}_1,\ldots,\underline{z}_{n+2} \in R^n$ and suppose that no n of them are linearly dependent. Then there are either 0 or 2 subsets $\Sigma \subset \{1,\ldots,n+2\}$ of cardinality $n+1$ such that the origin $\underline{0}$ is in the convex cover of the $\underline{z}_j$, $j \in \Sigma$.

Proof. Left to reader.

We now adopt a different definition of label. Instead of (4) we take

$$\phi(\underline{v}) \in T , \tag{16}$$

where $T \subset R^{n+1}$ is the hyperplane

$$\sum_{0 \le j \le n} x_j = 0 , \tag{17}$$

with the condition

$$v_i = 0 \Rightarrow (\phi(v))_i \ge 0 . \tag{18}$$

We say that a facet of our simplicial decomposition is well-labelled if $\underline{0}$ is in the convex cover of the labels of the vertices.

Lemma 4. There are infinitely many well-labelled facets.

Proof. It is left to the reader to verify that (18) implies that the facet S given by (12) is well-labelled (e.g. by Lemma 9 of Appendix A).

Let us first suppose (general position) that no n of the labels on any facet are linearly dependent. Then (18) implies that no facet which lies on a coordinate hyperplane can be well-labelled. By Lemma 3 every simplex has either 0 or 2 well-labelled facets. Hence, exactly as in the proof of Lemma 1, there is a sequence

$$S, S_1, S_2, \ldots \tag{19}$$

of distinct well-labelled facets.

If the labelling is not in general position, the above proof breaks down in several places. However, it is not difficult to see that every labelling is a limit (in an obvious sense) of labellings in general position: and so a standard compactness argument (diagonal process) gives a sequence (19). There is a more constructive variant of this argument. Let $\psi(\underline{v})$ be any labelling in general position and let t be a variable over R. We order the field $R(t)$ by $t > 0$ but $t < b$ for any $b > 0$ in R. Then $\phi(\underline{v}) + t\psi(\underline{v})$ is a labelling in general position over $R(t)$ in an obvious sense. Hence it gives rise to a unique sequence (19) of facets well-labelled for $\phi(\underline{v}) + t\psi(\underline{v})$ over $R(t)$. These facets are clearly also well-labelled for $\phi(\underline{v})$ over R, which concludes the proof of Lemma 4.

We now prove Kakutani's Theorem for S. Take

$$\phi(\underline{v}) = \underline{f} - \underline{s} \quad , \tag{20}$$

where $\underline{s} = \pi(\underline{v})$, as in (7), and where $\underline{f} \in F(\underline{s})$ is selected arbitrarily. Then the hypotheses of Lemma 4 are satisfied, and so, as in the proof of Brouwer's theorem, there are arbitrarily small simplexes with vertices $\underline{s}^{(0)},\ldots,\underline{s}^{(n)}$ (say) and with $\underline{f}^{(j)} \in F(\underline{s}^{(j)})$ such that $\underline{0}$ is in the convex cover of the $\underline{f}^{(j)} - \underline{s}^{(j)}$. By compactness and the hypothesis of upper semi-continuity, there is an $\underline{s}^* \in S$ and there are

$$\underline{f}^{*(j)} \in F(\underline{s}^*) \qquad (0 \le j \le n) \tag{21}$$

such that $\underline{0}$ is in the convex cover of the $\underline{f}^{*(j)} - \underline{s}^*$. But then $\underline{s}^* \in F(\underline{s}^*)$ by the hypothesis that $F(\underline{s})$ is convex for all $\underline{s}$. This is just the conclusion of Kakutani's theorem.

Reference

S. Kakutani. A generalization of Brouwer's fixed-point theorem. Duke Math. J. 8 (1941), 451-459.

APPENDIX C

Non-negative matrices

There is a rich theory of square matrices whose elements are non-negative real numbers, which was originated by Frobenius and Perron. We prove here only what we need.

Theorem 1. Let

$$A > 0 \tag{1}$$

be a square matrix. Then there is a $\mu = \mu(A)$ in

$$0 < \mu < \infty \tag{2}$$

with the following properties:

(i) μ is an eigenvalue of A and has an eigenvector $\underline{b} > \underline{0}$.

(ii) Every (real or complex) eigenvalue λ of A satisfies

$$|\lambda| \leq \mu \ . \tag{3}$$

(iii) For given ρ , a necessary and sufficient condition that there exists a $\underline{c} > 0$ with

$$A\underline{c} \geq \rho\underline{c} \tag{4}$$

is that

$$\rho \leq \mu \ .$$

(iv) For given $\sigma \geq 0$, a necessary and sufficient condition that there exist a $\underline{d} > \underline{0}$ with

$$\underline{d} >> \sigma A\underline{d} \tag{5}$$

is that

$$\mu\sigma < 1 \ .$$

(v) $(I - \sigma A)^{-1} > 0$ (6)

whenever

$$0 \leq \mu\sigma < 1 \ .$$

(vi) $\mu(A)$ is a nondecreasing function of each of the elements of A .

(vii) $\mu(A') = \mu(A)$, where A' is the transposed of A .

Proof. (iii) We define μ to be the supremum of the ρ for which there is a $\underline{c} > \underline{0}$ satisfying (4). Clearly (2) holds. Normalizing the $\underline{c}$ to lie on the simplex

$$S : \underline{x} > \underline{0} , \Sigma x_j = 1 , \quad (7)$$

we deduce by compactness that there is an $\underline{e} \in S$ with

$$A\underline{e} \geq \mu\underline{e} . \quad (8)$$

Then $\underline{c} = \underline{e}$ satisfies (4) for all $\rho \leq \mu$.

(vi) is now trivial.

(ii) Let $\underline{u}$ be an eigenvector to the eigenvalue λ, where $\underline{u},\lambda$ may be complex. Let $\underline{v} > \underline{0}$ be the vector consisting of the absolute values of the elements of $\underline{u}$. Clearly

$$A\underline{v} \geq |\lambda|\underline{v} ; \quad (9)$$

and (ii) follows from (iii).

We give two alternative proofs of (i).

(i) First proof. We must show that there is equality in (8) for some $\underline{e}$. Suppose, first, that for every index j there is some $\underline{e}$ for which either

$$(A\underline{e})_j > \mu e_j \quad (10_1)$$

or

$$(A\underline{e})_j = 0 . \quad (10_2)$$

On summing these $\underline{e}$ for $j = 1,\ldots,n$ we obtain an $\underline{e} = \underline{e}^*$ (say) which satisfies (10_1) or (10_2) for each j. But then

$$A\underline{e}^* \geq (\mu+\varepsilon)\underline{e}^* \quad (11)$$

for some $\varepsilon > 0$, contrary to the definition of μ. Hence we may suppose without loss of generality that

$$(A\underline{e})_1 = \mu e_1 \neq 0 \quad (12)$$

for every solution of (8).

By a compactness argument [e.g. considering only the $\underline{e}$ on (7)] there is a

$$\delta > 0 \quad (13)$$

such that every solution $\underline{e}$ of (8) satisfies

$$e_1 \geq \delta \sum_j e_j , \quad (14)$$

and there is an $\underline{e}'$ for which

$$e'_1 = \delta \sum_j e'_j \; . \qquad (15)$$

Put

$$\underline{e}'' = A\underline{e}' \; . \qquad (16)$$

Then

$$A\underline{e}'' - \mu\underline{e}'' = A(A\underline{e}' - \mu\underline{e}') \geq \underline{O} \; , \qquad (17)$$

so $\underline{e}''$ satisfies (8) . We have

$$e''_j \geq \mu e'_j \qquad (\text{all } j) \qquad (18)$$

by (8) and

$$e''_1 = \mu e'_1 \; . \qquad (19)$$

By (15) and (14) for $\underline{e} = \underline{e}''$ there must be equality in (18) for all j . Hence $\underline{b} = \underline{e}'$ is an eigenvector belonging to μ , as required.

(i) Alternative proof. Suppose, first, that

$$A >> O \; , \qquad (20)$$

and that there is inequality in (8) . Put $\underline{e}^* = A\underline{e}$. Then

$$A\underline{e}^* - \mu\underline{e}^* = \mu(A\underline{e} - \mu\underline{e}) > O \qquad (21)$$

and so, as in (11), we have a contradiction. Hence in case (20) there is always equality in (8) and we may take $\underline{b} = \underline{e}$.

Every A satisfying (1) is the limit of matrices for which (20) holds. The existence of $\underline{b}$ when (20) does not hold may now be deduced by compactness.

(vii) is immediate from (i) and (ii) .

There remain (iv) and (v). We give two alternative arguments. The first proves (v) first, and deduces (iv). The second starts with (iv).

(v) First proof. The infinite series

$$I + \sigma A + \ldots + (\sigma A)^k + \ldots \qquad (22)$$

converges when $\mu\sigma < 1$ by (ii). It sums to $(I-\sigma A)^{-1}$. The truth of (6) is now evident.

(iv) First proof. Suppose that $\mu\sigma < 1$ and let $\underline{z}$ be any vector with $\underline{z} >> \underline{O}$. Then

$$\underline{d} = (I-\sigma A)^{-1} \underline{z} > \underline{O} \qquad (23)$$

by (iv) and clearly satisfies (5).

Suppose, now, that (5) holds. By (vii) and (i) applied to the transposed of A there is a row vector $\underline{f} > \underline{O}$ such that $\underline{f}A = \mu\underline{f}$. Then (5) implies that

$$\underline{fd} > \sigma\mu\underline{fd} \quad ; \qquad (24)$$

and so $\sigma\mu < 1$.

(iv) Alternative proof. Apply Lemma 8, Corollary, of Appendix A to the matrix $D = I-\sigma A$. The (iv) follows immediately from (iii) for the transposed of A.

(v) Alternative proof. Suppose that $\mu\sigma < 1$. We show first that any real vector $\underline{g}$ with

$$\underline{g} > \sigma A\underline{g} \qquad (25)$$

necessarily satisfies $\underline{g} > \underline{O}$. For, if not, there is a $\theta \geq \underline{O}$ such that

$$\underline{d} + \theta\underline{g} > \underline{O} \text{ but not } \gg\underline{O} \qquad (26)$$

for the $\underline{d}$ given by (iv). Then (5),(25),(26) would give

$$\begin{aligned}\underline{d} + \theta\underline{g} &\gg \sigma A(\underline{d} + \theta\underline{g}) \\ &\geq \underline{O} \ , \end{aligned} \qquad (27)$$

in contradiction to (26).

The matrix $(I-\sigma A)^{-1}$ exists by (ii). The truth of (v) now follows by applying the preceding argument with $\underline{g}$ as each of the columns of $(I-\sigma A)^{-1}$ in turn.

We say that the matrix $A = (a_{ij})$ is reducible if there is some non-empty set S of indices such that

$$a_{ij} = 0 \qquad (i \in S \ , \ j \notin S) \ .$$

This is the same as saying that there is a permutation of the indices such that A takes the shape

$$A = \begin{pmatrix} A_{11} & A_{12} \\ 0 & A_{22} \end{pmatrix}$$

for submatrices A_{11}, A_{12}, A_{22}. If A is not reducible, then it is irreducible. We have now the

Corollary. Suppose that A is irreducible. Then

(viii) The eigenvector $\underline{b}$ given by (i) is unique (up to a multiplicative factor) and satisfies $\underline{b} \gg \underline{O}$.

(ix) Let $\underline{c} > \underline{0}$ be an eigenvector to eigenvalue λ. Then $\lambda = \mu$ and $\underline{c}$ is a multiple of $\underline{b}$.

<u>Proof</u>. (viii). Suppose it is not true that $\underline{b} \gg \underline{0}$. Let S be the set of indices k for which $b_k = 0$. Then $A\underline{b} = \mu\underline{b}$ implies that $a_{ij} = 0$ for $i \in S, j \notin S$, contrary to the assumption that A is irreducible.

Now let $\underline{a} > \underline{0}$ be another eigenvector belonging to μ. Then we can find $\theta \geq 0$ such that $\underline{a} - \theta\underline{b} \geq \underline{0}$ but not $\gg 0$. Then $\underline{a} - \theta\underline{b}$ is an eigenvector to μ, and there is a contradiction to what has just been proved unless $\underline{a} = \theta\underline{b}$.

(ix) By applying (viii) to the transposed of A, there is an $\underline{f} \gg \underline{0}$ with $\underline{f}A = \mu\underline{f}$. Then $\underline{fc} \neq 0$ and

$$\lambda\underline{fc} = \underline{fAc} = \mu\underline{fc} .$$

Hence $\lambda = \mu$ and (viii) applies to $\underline{b} = \underline{c}$.

Appendix C Exercises

1. Find an A and an $\underline{e}$ for which there is inequality in (8). [It is enough to consider 2×2 matrices.]
2. If A is irreducible, show that μ is an eigenvalue of multiplicity 1.
   [Hint. If not, there is a $\underline{c}$ linearly independent of $\underline{b}$ annihilated by $(A-\mu I)^2$, so $A\underline{c}-\mu\underline{c} = \theta\underline{b}$ for some θ. Use the technique of (ix) to show that $\theta = 0$.]
3. (a) If $A >> 0$, show that there is strict inequality in (3) for $\lambda \neq \mu$.
   (b) Show that the result of (a) does not extend to all irreducible A.
   [Hints. (a) $A-\delta I>0$ for some $\delta > 0$. Apply (3) with $A-\delta I$ for I. (b) It is enough to consider 2×2 matrices.]
4. Show that the adjoint
   $$\mathrm{Adj}(I-\sigma A) > 0 ,$$
   provided that $0 \leq \mu\sigma < 1$: and that this continues to hold for $\mu\sigma = 1$ provided that A is irreducible.
5. Let A be irreducible. For given index i let $T(i)$ be the set of $t > 0$ for which there is a set of indexes
   $$i = j_0, j_1, \ldots, j_t = i$$
   with
   $$a(j_r, r_{r+1}) \neq 0 \qquad (0 \leq r < t) ,$$
   where we write $a_{ij} = a(i,j)$ for typographical convenience. Show (i) that the greatest common divisor q of the $t \in T(i)$ is independent of i and (ii) that $T(i)$ contains all sufficiently large multiples of q.
   (a) Suppose that $q = 1$. Show that $A^N >> 0$ for all sufficiently large N. Deduce that there is

strict inequality in (3) for $\lambda \neq \mu$.

(b) Suppose that $q > 1$. Show that $n = qs$ for some integer s and that the indices $1,\ldots,n$ can be partitioned into q sets $S(1),\ldots,S(q)$ such that $a_{ij} = 0$ unless $i \in S(u)$, $j \in S(u+1)$ for some u , where u is taken modulo q . Show, further, that for all sufficiently large N the elements b_{ij} of A^{Nq} are non-zero precisely when i,j are in the same $S(u)$. Deduce that the eigenvalues λ of A with $|\lambda| = \mu$ are precisely the roots of $\lambda^q = \mu^q$.

6. Let A be irreducible, and suppose that not all the diagonal elements of A are 0 . Show that there is inequality in (3) for $\lambda \neq \mu$.
[*Hint*. Previous exercise.]

INDEX